Oxford Mathematics 6

Primary Years Programme

Contents

OXFORD
UNIVERSITY PRESS
AUSTRALIA & NEW ZEALAND

Practice

1 What is the value of the red digit in each number?

    a   734 815   b   62 759 377

    c   6 219 730   d   13 585 104

    e   5 487 800   f   218 819 999

2 Write the numbers from question 1 in words.

    a _____

    b _____

    c _____

    d _____

    e _____

    f _____

3 Expand these numbers. The first one has been started for you.

    a   413 629:

          400 000 + _____

    b   3 746 123:

          _____

    c   52 065 350:

          _____

    d   43 200 806:

          _____

    e   680 405 020:

          _____

Challenge

1 Look at the digit cards.

| 5 | 2 | 8 | 7 | 9 | 3 | 1 | 0 |

a What is the **smallest** whole number that can be made using all the cards if the 0 is in the millions place?

_____

b What is the **largest** number that can be made if the digit 1 is in the tens of millions place?

_____

c What is the **smallest** number that can be made if the 9 is the last digit and is used as nine tenths?

_____

d What is the **largest** number that can be made if the digits 7, 8 and 9 are used as decimals?

_____

2 Match each number to the correct spike abacus.

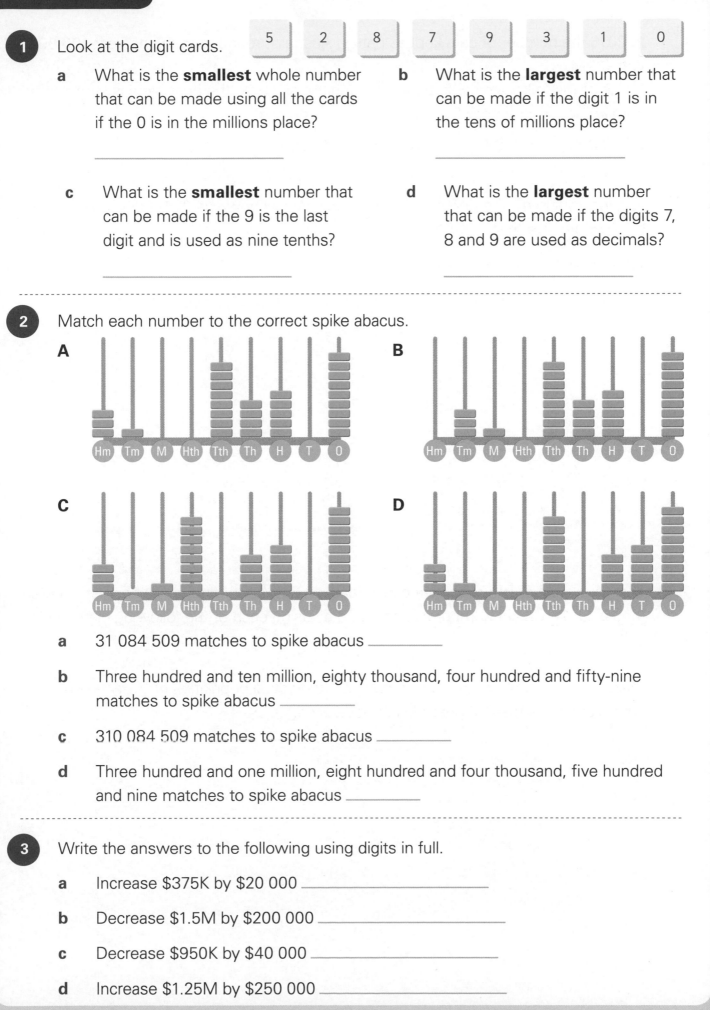

a 31 084 509 matches to spike abacus _____

b Three hundred and ten million, eighty thousand, four hundred and fifty-nine matches to spike abacus _____

c 310 084 509 matches to spike abacus _____

d Three hundred and one million, eight hundred and four thousand, five hundred and nine matches to spike abacus _____

3 Write the answers to the following using digits in full.

a Increase $375K by $20 000 _____

b Decrease $1.5M by $200 000 _____

c Decrease $950K by $40 000 _____

d Increase $1.25M by $250 000 _____

1 In 2016 the number of cats kept as pets in Australia, rounded to the nearest 100 000, was 3.9M.

Some of the following could be the exact number. Circle the possible answers.

3 487 245	3 878 224	4 900 010	3 995 110
3 949 999	3 898 100	3 918 885	4 000 001

2 In a recent survey, the dog population of China was shown as 22 908 545. This could be rewritten like this:

22 000 000 + 900 000 + 8000 + 545

Rewrite the dog population of China in as many ways as you can think of.

3 The population of Browntown recently reached 2.3M, rounded to the nearest hundred thousand. The exact number can be found by using each of the following digits once:

1   2   3   5   6   7   9

Make a list of numbers that could represent the actual population of Browntown. (There are more than 20 possible correct answers.)

OXFORD UNIVERSITY PRESS

Practice

1 Lightly shade all the square numbers on this hundred grid.

1	2	3	4	5	6	7	8	9	10
11	12	13	14	15	16	17	18	19	20
21	22	23	24	25	26	27	28	29	30
31	32	33	34	35	36	37	38	39	40
41	42	43	44	45	46	47	48	49	50
51	52	53	54	55	56	57	58	59	60
61	62	63	64	65	66	67	68	69	70
71	72	73	74	75	76	77	78	79	80
81	82	83	84	85	86	87	88	89	90
91	92	93	94	95	96	97	98	99	100

2 Underline the sentence that is a statement of fact:

All square numbers are odd.

All square numbers are even.

There is an equal number of odd and even square numbers.

There are more odd square numbers than even square numbers.

There are more even square numbers than odd square numbers.

3 Circle any of the options below that complete the following sentence correctly.

If you add together the first five odd numbers, the answer is the same as:

5×5 1×5 5^2 5×2 2×5

4 Circle the first 10 triangular numbers on the hundred grid in question 1.

5 Which numbers on the hundred grid are both triangular and square numbers?

_____

Challenge

1 The 11th square number is 121. The sum of which consecutive 11 odd numbers is 121?

_____

2 **a** What is 13^2? _____

b Write the addition sequence that will give the same answer as 13^2.

_____

3 The 10th triangular number is 55. The 11th triangular number can be found by adding what to 55?

_____

4 This pattern of beads shows the third triangular number, which is 6.

a Draw a pattern of beads showing the 11th triangular number.

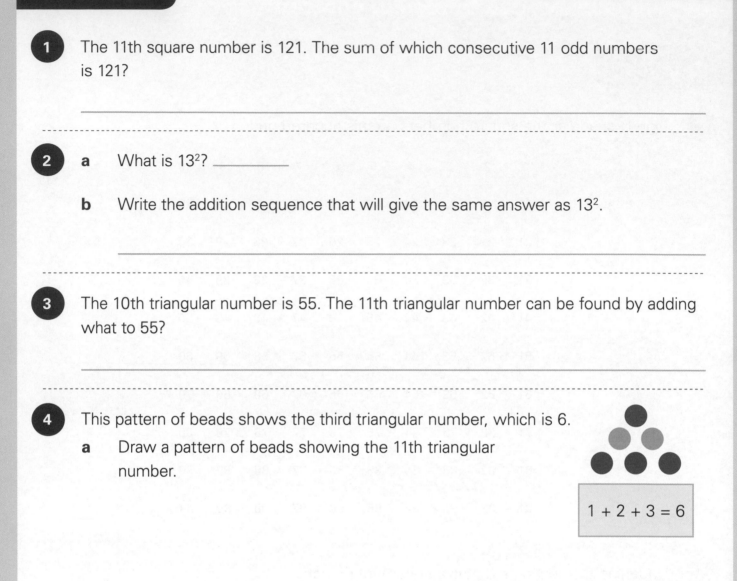

$1 + 2 + 3 = 6$

b Write the addition fact below the pattern using numbers that are added together to show the 11th triangular number.

OXFORD UNIVERSITY PRESS

1 Drawing and counting the dots in a small triangular number, such as 6, is simple. Larger triangular numbers can take a long time to work out in this way. However, there is a mathematical way of calculating **any** triangular number.

To understand this, we'll look at the second triangular number (2nd TN).

2nd TN

- Since it is the 2nd TN, we begin by drawing the dots as a right-angled triangle that is 2 dots high and 2 dots wide.

- Next, we **double the number of dots and make a rectangle**.

- The rectangle is one dot wider than it is high (2 dots high × (2 + 1) dots wide). We can write this more simply as:

The rectangle is 2 × (2 + 1) = 2 × 3 = 6 dots.

- We **doubled** the number of dots to make the rectangle, so we **halve** the number of dots to find the triangular number.

- The 2nd TN is 2 × (2 + 1) ÷ 2 = 2 × 3 ÷ 2 = 3 dots.

This will work for **any** triangular number. We used 2 for the 2nd TN, so we use 3 for the 3rd TN.

a Look at the rectangle and complete the calculation:

3rd TN = 3 × (3 + 1) ÷ 2

= _____

b Use the diagrams and the calculation to prove that the fifth triangular number is 15.

_____

_____

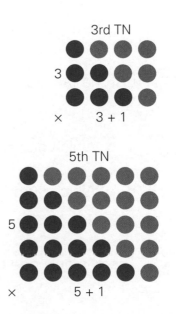

- -

2 Use the rule for calculating triangular numbers to work out as many triangular numbers as you wish on a separate piece of paper. You could begin by calculating the value of the 10th and 20th triangular numbers.

Practice

Remember: A prime number has just two factors—1 and itself. All other numbers are composite numbers

1 Complete the chart to show the factors of the prime and composite numbers between 21 and 30.

Number	Factors (numbers it can be divided by)	How many factors?	Is it prime or composite?
21			
22			
23			
24			
25			
26			
27			
28			
29			
30			

2 List the factors of 12. _____

3 a Which number between 21 and 30 has the same amount of factors as 12? _____

  b Which numbers between 21 and 30 have more factors than 12? _____

4 There is only one even prime number. What is it? _____

5 a How many factors does the square number on the chart above have? _____

  b Name a square number lower than 10 that has the same number of factors as the square number on the chart above. _____

OXFORD UNIVERSITY PRESS

Challenge

1 Fill in the gaps to show the prime factors of the following:

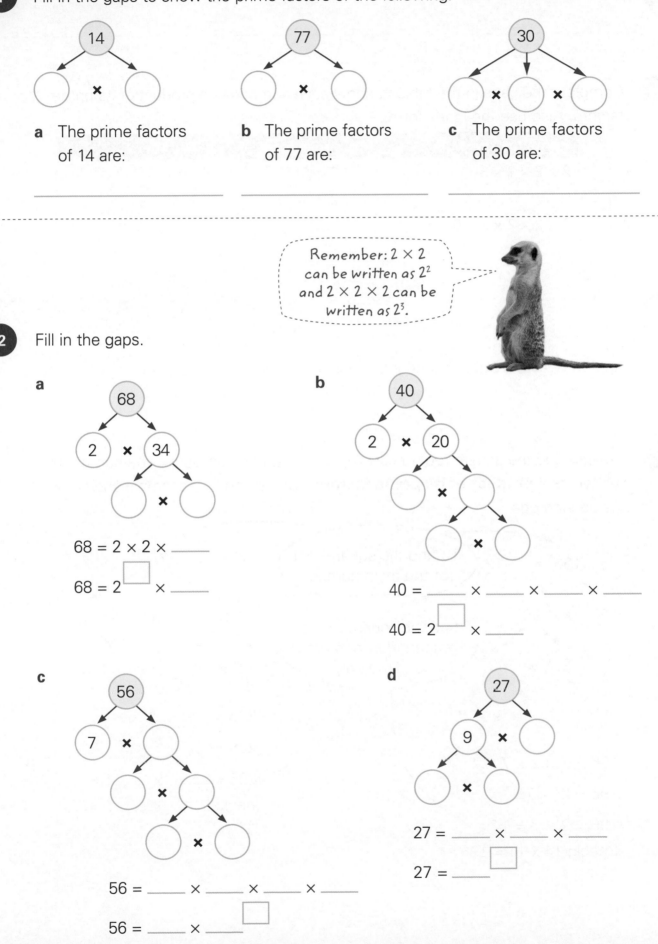

a The prime factors of 14 are:

_____

b The prime factors of 77 are:

_____

c The prime factors of 30 are:

_____

Remember: 2 × 2 can be written as 2^2 and 2 × 2 × 2 can be written as 2^3.

2 Fill in the gaps.

a

68 = 2 × 2 × _____

68 = $2^{\square}$ × _____

b

40 = _____ × _____ × _____ × _____

40 = $2^{\square}$ × _____

c

56 = _____ × _____ × _____ × _____

56 = _____ × _____$^{\square}$

d

27 = _____ × _____ × _____

27 = _____$^{\square}$

1 How do you know that 37 is a prime number?

_____

2 Complete the gaps in the table to show each number as a product of its prime factors. Also use the index form.

	Prime factors	Index form	Number
a	2 × 2 × 2 × 3	$2^3 \times 3$	
b		$2^3 \times 5^2$	200
c	2 × 2 × 2 × 2 × 5		
d			84

3 a _____^{2} = 16   b _____^{3} = 8   c _____^{2} = 49   d _____^{3} = 27

4 Another way of writing question 3a would be, "Find the square root of 16" or "√16".

a √36 = _____   b √64 = _____   c √81 = _____

5 We can find the square root of certain, less-common numbers if the number is written as a **product of its prime factors**. Look at the first example and complete the second one.

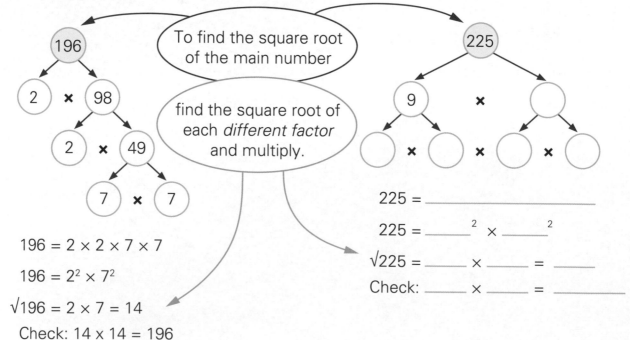

To find the square root of the main number

find the square root of each *different factor* and multiply.

196 = 2 × 2 × 7 × 7

196 = $2^2 \times 7^2$

√196 = 2 × 7 = 14

Check: 14 x 14 = 196

225 = _____

225 = _____^{2} × _____^{2}

√225 = _____ × _____ = _____

Check: _____ × _____ = _____

OXFORD UNIVERSITY PRESS

UNIT 1: TOPIC 4
Mental strategies for addition and subtraction

Practice

1 You could use the split strategy to solve these:

a   231 + 427 _____   b   216 + 443 _____

c   536 + 822 _____   d   1823 + 636 _____

e   4285 + 714 _____   f   4724 + 8265 _____

2 You could use the compensation (rounding) strategy to solve these:

Some of these involve trading.

a   638 – 399 _____

b   427 + 302 _____

c   362 + 698 _____

d   768 – 501 _____

e   857 – 290 _____

f   998 + 403 _____

3 Explain the method that you use to solve the following:

a   551 + 348 = _____

b   1472 – 299 = _____

c   4350 + 4350 = _____

d   24 746 – 520 = _____

Challenge

1 Here is a list of five of the most expensive paintings ever sold. Decide how to round the numbers. Explain how you have rounded them.

	Painting title	Price	Artist	Rounded price	Rounded to the nearest
a	*Salvator Mundi* (painted around 1500)	$450 300 000	Leonardo da Vinci		
b	*Les Femmes d'Alger* (painted in the 1950s)	$179 000 000	Pablo Picasso		
c	*Nu Couché* (painted in 1917/18)	$170 400 000	Amedeo Modigliani		
d	*Three Studies of Lucian Freud* (painted in the 1960s)	$142 400 000	Francis Bacon		
e	*The Scream* (painted in the 1890s)	$119 922 500	Edvard Munch		

2

a If the same person bought paintings **a** and **d**, would the amount be closer to $592M or $593M? _____

b How much less was painting **c** than painting **b**? _____

c How much less than $120M was painting **e**? _____

d What was the total paid for paintings **c** and **d**? _____

3 Estimate the answer to each problem below, then check on a calculator. Make sure your answer is close to the calculator answer. If it isn't, check what went wrong.

	Problem	Round the numbers	Estimate the answer	Calculator answer
e.g.	4195 + 3030	4000 + 3000	7000	7225
a	3085 + 1877			
b	6018 − 3119			
c	35 056 + 4023			
d	29 856 − 20 028			
e	400 247 + 49 588			
f	398 108 − 197 434			
g	5 498 858 − 2 889 917			

OXFORD UNIVERSITY PRESS

1 Two adults and two children go on a holiday. The flights cost $199 for each child and $299 for each adult. The accommodation costs $1299 and they spend $310 on food.

 a Is the price of the holiday closest to $2550, $2600, $2650 or $2700?

 _____

 b What is the exact price of the holiday? _____

2 You may need to use a calculator to solve this problem.

 a Audrey wrote the 9 digits for her friend like this:

 # 123456789

 She asked, "What would you need to add to the number to get this answer:

 # 987654321

 b Write the number that would need to be added. Write your answer in digits, with the correct spacing, and in words.

 _____

3 A theatre holds around 500 people. At the end of eight shows, exactly 4000 people have attended, but no two performances had exactly the same number in the audience. Show in the table how many people could have been at each performance.

Performance	Number of people
1	
2	
3	
4	
5	
6	
7	
8	
Total	

Practice

1 Set out these addition problems vertically. Look for patterns in the answers.

a 107 + 16

+ _____

b 387 + 69

+ _____

c 425 + 27 + 337

+ _____

d 784 + 354 + 96

+ _____

e 1786 + 10 528 + 31

+ _____

f 89 316 + 5892 + 28 248

+ _____

g 758 306 + 392 048 + 84 213

+ _____

h 8 235 946 + 428 637 + 1 665 736 + 2 015 359

+ _____

i 845 386 + 57 214 936 + 35 678 523 + 22 581 099 + 7 136 845

+ _____

Remember: Keep the digits in the correct columns.

OXFORD UNIVERSITY PRESS

Challenge

Did you know that in Brazil there are more than one and a half million kilometres of unpaved roads? In France, however, there are no unpaved roads!

This list shows the total length of paved and unpaved roads in 11 countries. The list is in order from the country with the greatest length of paved roads, to the country with the smallest length of paved roads.

	Country	Length of paved roads (km)	Length of unpaved roads (km)	Total length of roads (km)
1	USA	4 165 110	2 265 256	
2	India	1 603 705	1 779 639	
3	China	1 515 797	354 864	
4	France	951 220	0	
5	Japan	925 000	258 000	
6	Russia	738 000	133 000	
7	Spain	659 629	6 663	
8	Canada	415 600	626 700	
9	UK	388 008	0	
10	Australia	336 962	473 679	
11	Brazil	96 353	1 655 515	

1 **a** Without using a calculator, find the total length of roads for each country and then fill in the last column of the table.

b The order changes when the total length of roads has been calculated. Re-order the list, from the country with the greatest total length of all roads, to the country with the smallest total length of all roads.

1 The answer is 123 321. There are two 5-digit addends. Write three different sums that would give the correct answer.

2 Find four numbers that can be added together to give an answer of 123 456 789. None of the numbers should have less than four digits.

3 The answer is 12 046. Use the digits 1, 0, 5 and 7 to write the sum. You can use each digit more than once. You may wish to use a calculator.

4 The answer is 987 654 321. There are four addends: one has 9 digits; one has 8 digits; one has 7 digits and one has 6 digits. Write two different sums that would give the correct answer.

OXFORD UNIVERSITY PRESS

UNIT 1: TOPIC 6
Written strategies for subtraction

Practice

1 Look for a connection between each subtraction and the one before it.

a

	3	6	9	7	4	8
−	2	4	8	5	3	6

b

	3	1	8	2	5	9
−		8	5	9	3	6

c

	4	7	2	7	9	1
−	1	2	9	3	5	7

d

	7	2	2	8	6	1
−	2	6	8	3	1	6

e

	9	7	1	4	8	5
−	4	0	5	8	2	9

f

	7	9	2	3	9	8
−	1	1	5	6	3	1

g

	1	1	4	6	1	2	3
−		3	5	8	2	4	5

h

	1	3	8	1	3	5	6
−		4	8	2	3	6	7

i

	1	1	2	4	8	5	4
−		1	3	4	9	5	6

j

	1	1	2	8	1	6	8
−		2	4	9	3	8	1

2 Write an algorithm to show that 343 965 subtracted from another number gives an answer of 767 676.

1 These subtractions involve zeros in the top line. Look for a connection between each subtraction and the one before it.

a

	4	0	4	6	3	1
−		8	3	3	1	0

b

	7	2	0	3	0	5
−	2	8	7	8	7	3

c

	6	3	0	0	4	2	9	6
−	1	9	7	8	9	9	7	5

d

	8	2	0	5	0	6	0	7
−	2	7	7	2	5	1	7	5

e

	9	4	0	0	1	0	0	5
−	2	8	5	6	4	4	6	2

f

	8	0	0	0	3	0	0	0
−		3	4	5	5	3	4	6

g

	9	0	1	4	5	0	9	4
−		2	4	8	6	3	2	9

h

	1	0	0	2	0	0	0	8	0
−			1	4	3	0	2	0	4

i

	5	0	0	5	0	0	5	0	0
−	3	7	7	0	4	6	1	7	9

j

	8	0	8	0	8	0	8	0	8
−	2	6	4	8	6	8	4	6	3

2 Write a 9-digit subtraction for which the answer is 123 456 789. The number on the top line of the subtraction should have 9 digits and include 6 zeros.

OXFORD UNIVERSITY PRESS

Mastery

1 In Australia, 40 731 tonnes of avocados were grown in 2007. By how much did this amount increase from the 33 825 tonnes that were produced in 2004?

_____

2 Over 40 000 tonnes of Australian avocados seems like a lot, but it is not much compared to the top five avocado-producing countries.

This table compares the amount of avocados each of the top five countries produced in 2007 and in 2004.

Country	Avocado production in tonnes (2007)	Avocado production in tonnes (2004)
Mexico	1 142 892 t	987 000 t
Chile	250 000 t	160 000 t
Indonesia	201 635 t	221 774 t
Colombia	193 996 t	170 985 t
Dominican Republic	183 468 t	218 790 t

a For which countries in the table did the production of avocados increase between 2004 and 2007? _____

_____

b By how many tonnes did the Dominican Republic's avocado production decrease between 2004 and 2007? _____

_____

c What was the difference between the amounts produced in Indonesia in 2004 and 2007? _____

_____

d What was the difference between the amounts produced in the top country and in Australia in 2004? _____

_____

e By how many tonnes was the amount produced in the top country greater than the total amount produced by the other four countries in 2007?

_____

_____

f By how much was the total produced by the five countries greater in 2007 than it was in 2004? _____

_____

Practice

1 Fill in the table to multiply by 10, by 100 and by 1000.

		× 10	× 100	× 1000
a	43			
b	87			
c	235			
d	$7.75			
e	250			
f	36			
g	$0.25			
h	$10.15			

Remember: When you multiply by 10 the digits move one place bigger. When you divide by 10 the digits move one place smaller.

2 Fill in the table to divide by 10 and then write the multiplication-fact partner. There are two reminders for you about using mental strategies to divide by 10.

		÷ 10	Write the multiplication-fact partner
e.g.	230	23	23 × 10 = 230
e.g.	25	2.5	2.5 × 10 = 25
a	350		
b	3600		
c	5000		
d	$12.50		
e	73		

3 Fill in the table to divide by 100 and then write the multiplication-fact partner. There are two reminders for you about using mental strategies to divide by 100.

		÷ 100	Write the multiplication-fact partner
e.g.	900	9	9 × 100 = 900
e.g.	$325	$3.25	$3.25 × 100 = $325
a	800		
b	$255		
c	7000		
d	15 000		
e	7250		

OXFORD UNIVERSITY PRESS

Challenge

1 Use the doubles strategy to multiply by multiples of 10.

		× 10	× 20	× 40	× 80
e.g.	24	240	480	960	1920
a	11				
b	14				
c	21				
d	35				
e	60				

2 Use halving strategies to divide by multiples of 10.

		÷ 10	÷ 20	÷ 40	÷ 80
e.g.	1200	120	60	30	15
a	800				
b	4000				
c	720				
d	12 000				
e	4800				

3 Use a time-saving mental strategy to solve the following. Be ready to explain how you got the answers.

The short cut that works for you is the best short cut!

a 25 × 20 _____

b 24 × 40 _____

c 15 × 50 _____

d 14 × 40 _____

e 25 × 30 _____

f 125 × 20 _____

g 25 × 60 _____

h 42 × 40 _____

i 19 × 20 _____

j $1.30 × 50 _____

k $2.75 × 20 _____

l 448 ÷ 4 _____

m 450 ÷ 5 _____

n 1800 ÷ 5 _____

4 Explain a mental strategy that you would use to multiply 22 × 14.

_____

_____

1 If you know that 9 × 7 = 63, you also know that 90 × 70 = (Circle one):

630   6030   6300   63 000

2 **a** 600 × 40 = 24 000. What is the connection between the number of zeros in the two numbers being multiplied and in the answer?

_____

b Test whether the connection you noticed works for other numbers.

3 **a** Circle the sentences that would give the correct answer to 648 ÷ 20:

- You could halve 648 and move the digits one place smaller.
- You could divide 648 by 10 and then halve the answer.
- You could halve 648 and move the digits one place bigger.
- You could divide 648 by 2 and then by 10.

b Choose a strategy to divide 648 by 20. Explain why you chose that strategy.

4 A school needs 30 new outdoor seats. The first supplier sells five seats for $107.50. A second offers a crate of 32 seats for $624. If you were the principal, which deal would you choose? Give a reason for your answer.

UNIT 1: TOPIC 8
Written strategies for multiplication

Practice

1 Use short or extended multiplication to solve the following:

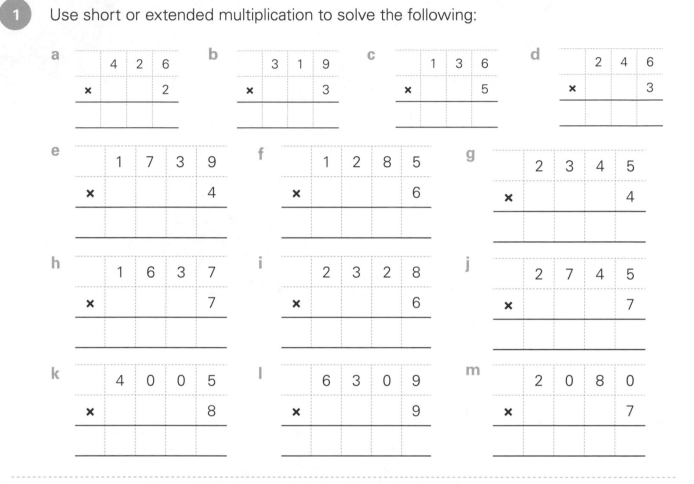

a
```
    4  2  6
×         2
─────────────
```

b
```
    3  1  9
×         3
─────────────
```

c
```
    1  3  6
×         5
─────────────
```

d
```
    2  4  6
×         3
─────────────
```

e
```
  1  7  3  9
×          4
─────────────
```

f
```
  1  2  8  5
×          6
─────────────
```

g
```
  2  3  4  5
×          4
─────────────
```

h
```
  1  6  3  7
×          7
─────────────
```

i
```
  2  3  2  8
×          6
─────────────
```

j
```
  2  7  4  5
×          7
─────────────
```

k
```
  4  0  0  5
×          8
─────────────
```

l
```
  6  3  0  9
×          9
─────────────
```

m
```
  2  0  8  0
×          7
─────────────
```

2 Six people go on a trip. The cost for each is $3078. How much do they pay altogether?

3 A builder sells seven new houses. The price for the first six is $349 845 each. The last one is $12 000 cheaper. How much does the builder get for all seven houses?

Challenge

Remember: Putting in the zero moves everything over one place for you.

1 Multiplying by a multiple of 10

a
```
    2 6
×   2 0
_____
      0
```

b
```
    3 4
×   2 0
_____
```

c
```
    2 7
×   3 0
_____
```

d
```
    5 3
×   4 0
_____

_____
```

e
```
    2 7
×   6 0
_____

_____
```

f
```
    4 4
×   5 0
_____

_____
```

g
```
  1 7 3
×   3 0
_____

_____
```

h
```
  2 5 3
×   6 0
_____

_____
```

i
```
  6 9 4
×   7 0
_____

_____
```

j
```
1 6 2 8
×     5 0
_____

_____
```

k
```
2 5 9 5
×     6 0
_____

_____
```

l
```
2 5 8 3
×     9 0
_____

_____
```

2 Multiplying by two digits

Multiplying by two digits is like doing two multiplications in one.

a
```
      2 4
×     2 5
_____       ← 24 × 5
.........
+       0       ← 24 × 20
_____

_____
```

b
```
      3 5
×     2 6
_____
.........
+
_____

_____
```

c
```
      4 8
×     1 8
_____
.........
+
_____

_____
```

d
```
    1 5 8
×     2 4
_____
.........
+
_____

_____
```

e
```
    3 0 8
×     6 5
_____
.........
+
_____

_____
```

f
```
    2 9 7
×     4 8
_____
.........
+
_____

_____
```

OXFORD UNIVERSITY PRESS

Mastery

1 Did you know that the five largest sea animals are all whales?

Use the lengths and masses of the various whales in the table to answer the questions below.

a Which type of whale has a mass that is closest to twice that of another whale?

_____

b What would the total mass of 23 northern right whales be?

_____

c If 15 blue whales were put nose to tail, what would the total length be? _____

	Mammal	Length	Mass
1	Blue whale	33.5 m	137 000 kg
2	Bowhead whale	20.2 m	86 000 kg
3	Northern right whale	18.35 m	77 700 kg
4	Fin whale	25.5 m	63 400 kg
5	Sperm whale	18.35 m	43 700 kg

d Which would have a greater mass, 26 fin whales or 12 blue whales? _____

e 15 northern right whales (nose to tail) would stretch the same distance as 15 sperm whales, but how much heavier would they be? _____

2 When some numbers are multiplied together they start to make an interesting pattern. Square numbers can make an interesting pattern, but you may need a calculator to find them. For example, 111 × 111 makes a pattern that starts with 123 … Try it for yourself, then continue the pattern. Your calculator might not be big enough, but if you get all the way to the square of 111 111 111, the answer is 12 345 678 987 654 321.

If you square 101 010 the answer is ten, twenty, thirty, twenty, ten, zero—or ten billion, two hundred and three million, twenty thousand, one hundred (10 203 020 100).

Use a calculator and spare paper to investigate patterns with multiplications. For example, you've seen that 11 × 11 = 121. Now try 111 × 11. And 1111 × 11. What other patterns can you find? Don't forget to write down the answers for others to see.

Practice

1 Complete the following, writing the reminder in two ways.

> The remainder for $25 \div 4$ can be written as 6 r1 or $6\frac{1}{4}$.

a $13 \div 4$ _____ b $31 \div 4$ _____

c $33 \div 8$ _____ d $64 \div 9$ _____

e $75 \div 8$ _____ f $59 \div 7$ _____

g $74 \div 9$ _____ h $59 \div 6$ _____

2 Complete each algorithm, showing the remainder in two ways.

a $4 \overline{)5\ 6\ 7}$ $4 \overline{)5\ 6\ 7}$ b $5 \overline{)3\ 7\ 2}$ $5 \overline{)3\ 7\ 2}$

c $6 \overline{)2\ 0\ 6}$ $6 \overline{)2\ 0\ 6}$ d $7 \overline{)7\ 4\ 5}$ $7 \overline{)7\ 4\ 5}$

e $8 \overline{)2\ 5\ 7\ 0}$ $8 \overline{)2\ 5\ 7\ 0}$ f $9 \overline{)1\ 7\ 5\ 4}$ $9 \overline{)1\ 7\ 5\ 4}$

g $4 \overline{)4\ 1\ 3\ 9}$ $4 \overline{)4\ 1\ 3\ 9}$ h $7 \overline{)3\ 3\ 0\ 9}$ $7 \overline{)3\ 3\ 0\ 9}$

3 Re-write using the $\overline{)}$ symbol. Write the remainder using the "r_" format.

a $1286 \div 5$ b $1844 \div 3$ c $2277 \div 6$

d $5826 \div 9$ e $8046 \div 4$ f $7055 \div 7$

OXFORD UNIVERSITY PRESS

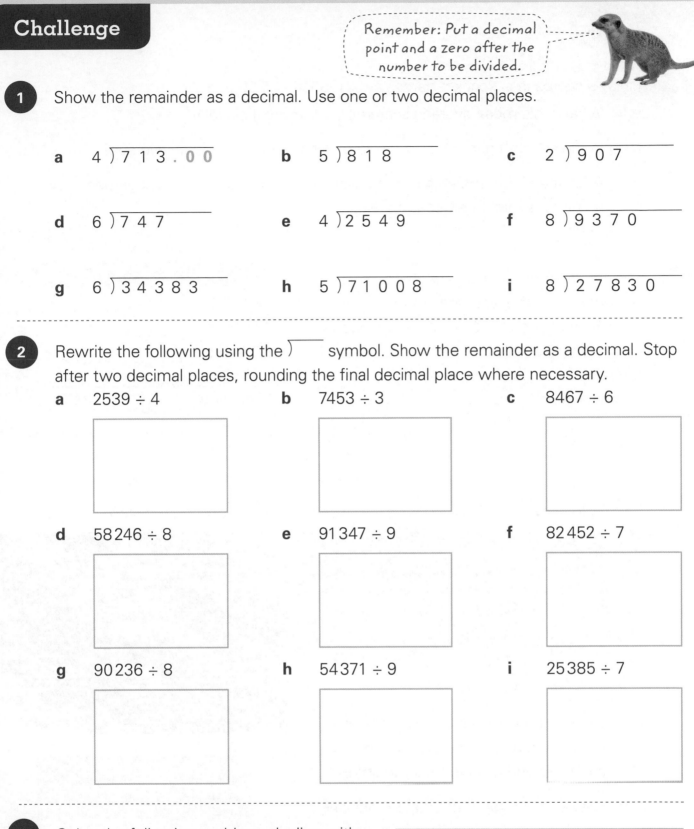

Challenge

Remember: Put a decimal point and a zero after the number to be divided.

1 Show the remainder as a decimal. Use one or two decimal places.

a 4) 7 1 3 . 0 0

b 5) 8 1 8

c 2) 9 0 7

d 6) 7 4 7

e 4) 2 5 4 9

f 8) 9 3 7 0

g 6) 3 4 3 8 3

h 5) 7 1 0 0 8

i 8) 2 7 8 3 0

2 Rewrite the following using the) symbol. Show the remainder as a decimal. Stop after two decimal places, rounding the final decimal place where necessary.

a 2539 ÷ 4

b 7453 ÷ 3

c 8467 ÷ 6

d 58 246 ÷ 8

e 91 347 ÷ 9

f 82 452 ÷ 7

g 90 236 ÷ 8

h 54 371 ÷ 9

i 25 385 ÷ 7

3 Solve the following problem, dealing with the remainder in the most appropriate way. A machine makes 23 455 pencils in a day. Each box holds eight pencils. How many boxes of pencils are made?

1 Finlay's scores in six cricket games are 21, 85, 4, 47, 59 and 101.

    a What is his mean average score (to two decimal places)? _____

    b In the seventh game he scores zero. What is his new mean average? _____

    c After the eighth match his mean average is higher than after six games. How many runs might he get in the eighth game? _____

2 The TV weather reporter said that the average temperature in the last week of the school holidays had been 31°C. The table shows the actual temperatures for three of the days. Fill in the table to show what the other temperatures might have been.

Day	Temperature
Sunday	
Monday	31°C
Tuesday	
Wednesday	
Thursday	34°C
Friday	29.5°C
Saturday	

3 This table shows the four countries in which the most potatoes are eaten.

Country	Amount of potatoes per person per week	Amount of potatoes per person per day
Belarus	3420.41 g	
Poland	2616.11 g	
Russia	2513.91 g	
Ireland	2436.07 g	

    a Work out the amount of potatoes eaten per person per day in each of the countries and fill in the final column.

    b Research the top four potato-producing countries, and record your findings in the table. Do you think these countries will be the same as those in the table above?

Country	Potato production per year in tonnes

OXFORD UNIVERSITY PRESS

Practice

1 Show the following operations on the number lines. Write the number sentence for each operation. The first number line has been started for you.

a Increase −4 by 5.

−8 −7 −6 −5 −4 −3 −2 −1 0 1 2 3 4 5 6 7 8

Number sentence: _____

b Decrease 3 by 5.

−8 −7 −6 −5 −4 −3 −2 −1 0 1 2 3 4 5 6 7 8

Number sentence: _____

c Decrease 6 by 8.

−8 −7 −6 −5 −4 −3 −2 −1 0 1 2 3 4 5 6 7 8

Number sentence: _____

d Decrease 0 by 7.

−8 −7 −6 −5 −4 −3 −2 −1 0 1 2 3 4 5 6 7 8

Number sentence: _____

e Increase − 6 by 9.

−8 −7 −6 −5 −4 −3 −2 −1 0 1 2 3 4 5 6 7 8

Number sentence: _____

f Increase − 2 by 7.

−8 −7 −6 −5 −4 −3 −2 −1 0 1 2 3 4 5 6 7 8

Number sentence: _____

g Decrease 1 by 7.

−8 −7 −6 −5 −4 −3 −2 −1 0 1 2 3 4 5 6 7 8

Number sentence: _____

h Increase 2 by 5.

−8 −7 −6 −5 −4 −3 −2 −1 0 1 2 3 4 5 6 7 8

Number sentence: _____

i Increase 0 by 6.

−8 −7 −6 −5 −4 −3 −2 −1 0 1 2 3 4 5 6 7 8

Number sentence: _____

j Decrease 7 by 11.

−8 −7 −6 −5 −4 −3 −2 −1 0 1 2 3 4 5 6 7 8

Number sentence: _____

k Increase −7 by 11.

−8 −7 −6 −5 −4 −3 −2 −1 0 1 2 3 4 5 6 7 8

Number sentence: _____

l Increase 0 by 8.

−8 −7 −6 −5 −4 −3 −2 −1 0 1 2 3 4 5 6 7 8

Number sentence: _____

2 What would a calculator show if you pressed the following keys?

a $5 - 10 =$ _____

b $-10 + 11 =$ _____

c $25 - 50 =$ _____

d $30 - 100 =$ _____

e $-75 + 100 =$ _____

f $57 - 60 =$ _____

1 Find the counting number for the number lines, then fill in the missing numbers.

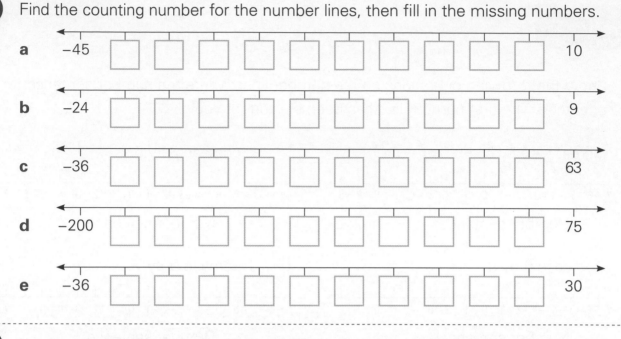

a −45 [ ] [ ] [ ] [ ] [ ] [ ] [ ] [ ] [ ] [ ] 10

b −24 [ ] [ ] [ ] [ ] [ ] [ ] [ ] [ ] [ ] [ ] 9

c −36 [ ] [ ] [ ] [ ] [ ] [ ] [ ] [ ] [ ] [ ] 63

d −200 [ ] [ ] [ ] [ ] [ ] [ ] [ ] [ ] [ ] 75

e −36 [ ] [ ] [ ] [ ] [ ] [ ] [ ] [ ] [ ] [ ] 30

2 It may surprise you to learn that, as well as recording temperatures of over 40°C, every state and territory in Australia has had below freezing temperatures!

Use the information in the table to complete the tasks below.

State or territory	Highest temp.	Year	Lowest temp.	Year
Australian Capital Territory	42.2°C	1968	−10°C	1971
New South Wales	50°C	1939	−23°C	1994
Northern Territory	48.3°C	1960	−7.5°C	1976
Queensland	49.5°C	1972	−11°C	1895
South Australia	50.7°C	1960	−8.2°C	1976
Tasmania	40.8°C	1976	−13°C	1983
Victoria	47.2°C	1939	−12.8°C	1947
Western Australia	50.5°C	1998	−6.7°C	1969

a In which state or territory was there a difference of 60.5° between the lowest and highest temperatures? _____

b If the lowest recorded temperature in the list were 8° warmer, what would the temperature be? _____

c What is the difference between the lowest and highest temperatures in the Northern Territory? _____

d What is the largest difference between the lowest and highest temperatures, and in which state or territory? _____

Mastery

1   **a**   Fill in the gaps on this part of an imaginary bank statement.

Date	Credit $ (Money going in)	Debit $ (Money going out)	Balance $
1 June	500.00		500.00
3 June		125.50	
5 June			305.50
10 June		420.10	
16 June		81.75	
21 June	325.00		
30 June			−220.30

b   If $500 were credited on 1 July, what would the new balance be? _____

---

2   Here are some rules for a board game called *Opposites*.

- Players start at zero.
- Player A tosses a coin marked *Positive* and *Negative* and then rolls a dice.
- Player A moves forwards (**+**) or backwards (**–**) along a number line according to the number on the dice.
- Player B moves a number of places that is **opposite** to Player A's number.
- After five turns, the player with the higher number wins.

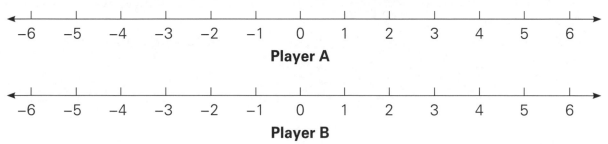

Player A

Player B

The table below shows the results of the five throws.

a   Complete the table to show the number each player moves using + and – to show the direction.

b   Plot the players' moves on the number lines.

c   Who is the winner and by how much? _____

Throw	+ or −	Number on dice	Player A moves	Player B moves
1	negative	2	−2	
2	positive	3		
3	negative	4		
4	negative	1		
5	positive	6		

Practice

1 Circle the expression with the same value as the exponent on each line.

a	3^3	$3 + 3 + 3$	3×3	$3 \times 3 \times 3$
b	5^4	$5 \times 5 \times 5 \times 5$	$5 + 5 + 5 + 5$	5×4
c	6^2	6×2	6×6	$6 + 6$
d	4^4	4×4	$4 + 4 + 4 + 4$	$4 \times 4 \times 4 \times 4$

2 Fill in the gaps.

	Base number and exponent	Number of times the base number is used in a multiplication	Multiplication	Value of the number
a	7^2			
b			$8 \times 8 \times 8$	
c		two times		4
d			$10 \times 10 \times 10 \times 10 \times 10$	
e	7^3			

3 Fill in the gaps.

	Square number	Exponent	Square root fact
a	36	6^2	$\sqrt{36} = 6$
b	16		
c		5^2	
d			$\sqrt{49} = 7$

How are the exponents and the square root facts related?

OXFORD UNIVERSITY PRESS

Challenge

1 Edel designed a machine that could make 5 donuts per minute.

   a Use an exponent to show how many donuts she would make in 5 minutes.

     _____

   b Edel makes 5 donut machines. Use an exponent to show how many donuts she could make in 5 minutes now. _____

2 Gopal gets 4 'likes' for his cat video.

   a Use an exponent to show how many 'likes' Gopal has if each of the 4 people has 4 friends who 'like' the video too. _____

   b The 4 friends each have 4 friends who 'like' the video. Use an exponent to show how many 'likes' Gopal has now. _____

   c Those 4 friends then each have 4 friends who 'like' the video. Use an exponent to show how many 'likes' Gopal has now. _____

   d Those 4 friends also have 4 friends each who 'like' Gopal's video. Use an exponent to show how many 'likes' he has now. _____

3 Salim is making a square vegetable patch. Use approximate square roots to find out how long the side lengths of his patch will be if the area is:

   a 62 m^2 _____   **b** 29 m^2 _____

   c 75 m^2 _____   **d** 22 m^2 _____

4 Evelyn was making a list of numbers with exact square roots, but she got confused in her calculations. Help her out by circling all the numbers that have an exact square root. You may use a calculator to help.

121		529		1147
			196	1024
143	484	625		
	400		684	
		906		2098
289	1089		2116	

1   Sasha the scientist discovered a germ that doubled in number every day. On the first day, there were 2 germs. On the second day, there were 2×2 germs and so on. Use exponents to show how many germs there were as time went on across at least 5 different days of your choice.

2   Mr Sass the sport teacher wanted to make a square four-square court. If the total area he had available was 200 m^2, what size could he have made the court and what would the side lengths be? Draw and label at least three options.

3   Which of your options in question 2 do you think is the best one? Why?

OXFORD UNIVERSITY PRESS

Practice

This fraction wall can help you to find some equivalent fractions.

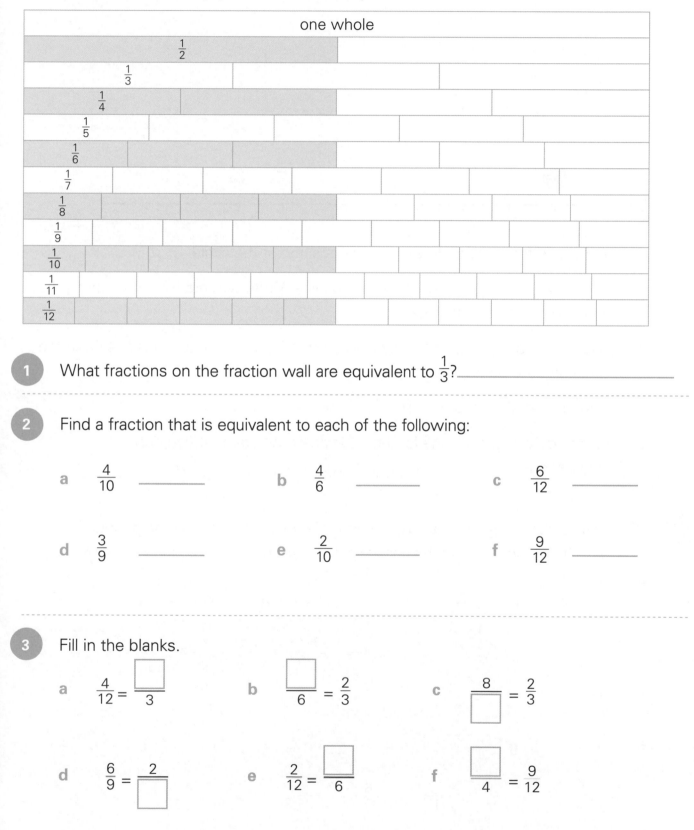

1. What fractions on the fraction wall are equivalent to $\frac{1}{3}$?_____

2. Find a fraction that is equivalent to each of the following:

   a $\frac{4}{10}$ _____

   b $\frac{4}{6}$ _____

   c $\frac{6}{12}$ _____

   d $\frac{3}{9}$ _____

   e $\frac{2}{10}$ _____

   f $\frac{9}{12}$ _____

3. Fill in the blanks.

   a $\frac{4}{12} = \frac{\square}{3}$

   b $\frac{\square}{6} = \frac{2}{3}$

   c $\frac{8}{\square} = \frac{2}{3}$

   d $\frac{6}{9} = \frac{2}{\square}$

   e $\frac{2}{12} = \frac{\square}{6}$

   f $\frac{\square}{4} = \frac{9}{12}$

Challenge

1 Divide and shade the shapes to show that:

a $\frac{2}{3}$ is equivalent to $\frac{4}{6}$

b $\frac{3}{4}$ is equivalent to $\frac{9}{12}$

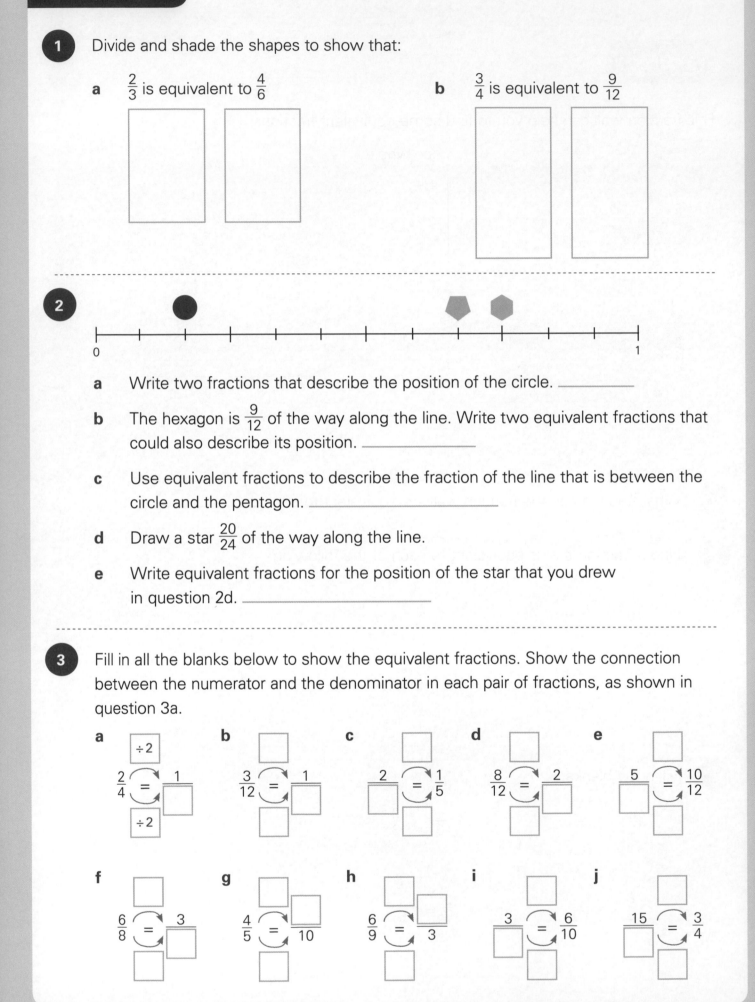

2

a Write two fractions that describe the position of the circle. _____

b The hexagon is $\frac{9}{12}$ of the way along the line. Write two equivalent fractions that could also describe its position. _____

c Use equivalent fractions to describe the fraction of the line that is between the circle and the pentagon. _____

d Draw a star $\frac{20}{24}$ of the way along the line.

e Write equivalent fractions for the position of the star that you drew in question 2d. _____

3 Fill in all the blanks below to show the equivalent fractions. Show the connection between the numerator and the denominator in each pair of fractions, as shown in question 3a.

a $\div 2$ $\frac{2}{4} = \frac{1}{\Box}$ $\div 2$

b $\frac{3}{12} = \frac{1}{\Box}$

c $\frac{2}{\Box} = \frac{1}{5}$

d $\frac{8}{12} = \frac{2}{\Box}$

e $\frac{5}{\Box} = \frac{10}{12}$

f $\frac{6}{8} = \frac{3}{\Box}$

g $\frac{4}{5} = \frac{\Box}{10}$

h $\frac{6}{9} = \frac{\Box}{3}$

i $\frac{3}{\Box} = \frac{6}{10}$

j $\frac{15}{\Box} = \frac{3}{4}$

OXFORD UNIVERSITY PRESS

Mastery

1 Reduce these fractions to their lowest equivalent form.

a $\dfrac{3}{15}$ _____

b $\dfrac{5}{20}$ _____

c $\dfrac{18}{24}$ _____

d $\dfrac{18}{36}$ _____

e $\dfrac{25}{100}$ _____

f $\dfrac{45}{50}$ _____

2 The shading on the circles below shows how much pizza is left over after a party.

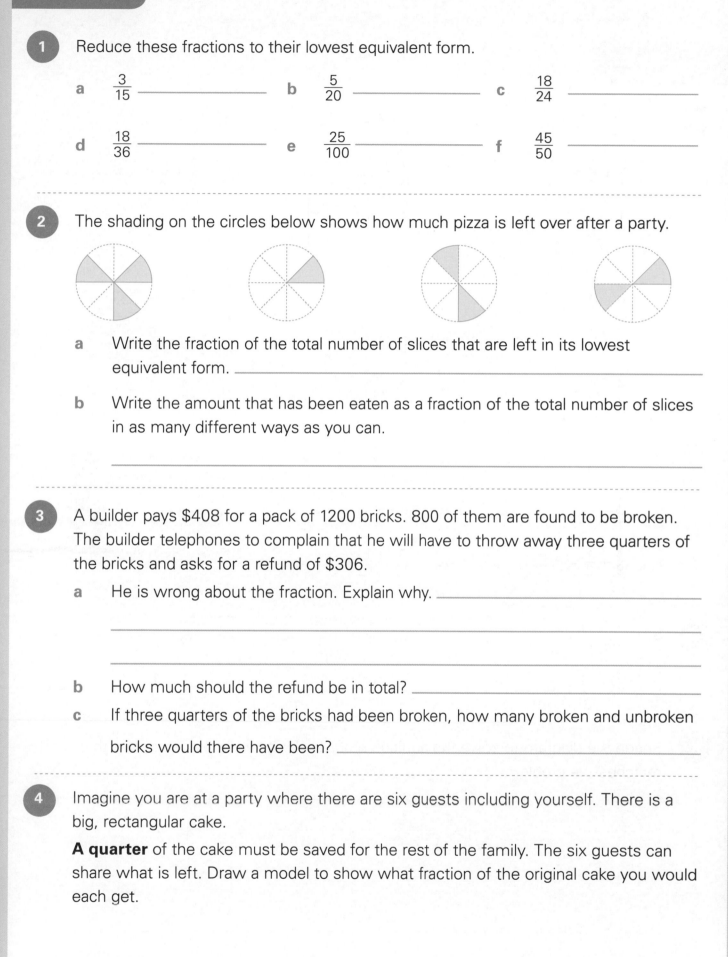

a Write the fraction of the total number of slices that are left in its lowest equivalent form. _____

b Write the amount that has been eaten as a fraction of the total number of slices in as many different ways as you can.

_____

3 A builder pays $408 for a pack of 1200 bricks. 800 of them are found to be broken. The builder telephones to complain that he will have to throw away three quarters of the bricks and asks for a refund of $306.

a He is wrong about the fraction. Explain why. _____

_____

b How much should the refund be in total? _____

c If three quarters of the bricks had been broken, how many broken and unbroken bricks would there have been? _____

4 Imagine you are at a party where there are six guests including yourself. There is a big, rectangular cake.

A quarter of the cake must be saved for the rest of the family. The six guests can share what is left. Draw a model to show what fraction of the original cake you would each get.

Practice

1

a $\dfrac{3}{10} + \dfrac{4}{10} =$ _____

b $\dfrac{5}{8} + \dfrac{2}{8} =$ _____

c $\dfrac{1}{5} + \dfrac{2}{5} =$ _____

d $\dfrac{3}{7} + \dfrac{3}{7} =$ _____

e $\dfrac{7}{10} + \dfrac{2}{10} =$ _____

f $\dfrac{2}{5} + \dfrac{2}{5} =$ _____

g $\dfrac{1}{8} + \dfrac{4}{8} =$ _____

h $\dfrac{4}{10} + \dfrac{6}{10} =$ _____

i $\dfrac{3}{8} + \dfrac{5}{8} =$ _____

2

a $\dfrac{5}{8} - \dfrac{3}{8} =$ _____

b $\dfrac{7}{10} - \dfrac{3}{10} =$ _____

c $\dfrac{9}{12} - \dfrac{5}{12} =$ _____

d $\dfrac{7}{8} - \dfrac{2}{8} =$ _____

e $\dfrac{3}{5} - \dfrac{1}{5} =$ _____

f $\dfrac{7}{12} - \dfrac{5}{12} =$ _____

g $\dfrac{4}{7} - \dfrac{4}{7} =$ _____

h $\dfrac{18}{20} - \dfrac{5}{20} =$ _____

i $\dfrac{14}{15} - \dfrac{4}{15} =$ _____

3

a $\dfrac{1}{8} + \dfrac{1}{4} =$ _____

b $\dfrac{1}{6} + \dfrac{2}{3} =$ _____

c $\dfrac{1}{5} + \dfrac{3}{10} =$ _____

d $\dfrac{1}{4} + \dfrac{5}{8} =$ _____

e $\dfrac{4}{9} + \dfrac{1}{3} =$ _____

f $\dfrac{3}{12} + \dfrac{1}{6} =$ _____

> Remember: Find a common denominator when adding and subtracting unlike fractions.

4

a $\dfrac{7}{10} - \dfrac{1}{5} =$ _____

b $\dfrac{7}{10} - \dfrac{1}{2} =$ _____

c $\dfrac{7}{12} - \dfrac{1}{2} =$ _____

d $\dfrac{9}{10} - \dfrac{1}{5} =$ _____

e $\dfrac{3}{4} - \dfrac{1}{12} =$ _____

f $\dfrac{3}{10} - \dfrac{1}{5} =$ _____

5 Shade the diagram to solve the addition problem. Write a number sentence that matches the problem.

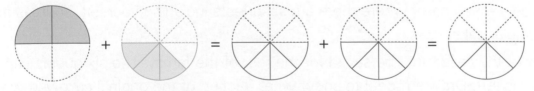

_____

OXFORD UNIVERSITY PRESS

Challenge

1 Find a common denominator to solve the following:

a $\dfrac{3}{8} + \dfrac{3}{4} =$ _____

b $\dfrac{2}{5} + \dfrac{9}{10} =$ _____

c $1\dfrac{3}{4} + \dfrac{5}{8} =$ _____

d $2\dfrac{2}{3} + \dfrac{5}{6} =$ _____

e $3\dfrac{7}{8} + \dfrac{3}{4} =$ _____

f $2\dfrac{5}{6} + \dfrac{1}{3} =$ _____

---

2 Remember to find a common denominator when solving the following:

a $1\dfrac{1}{8} - \dfrac{1}{4} =$ _____

b $3\dfrac{7}{10} - \dfrac{1}{2} =$ _____

c $2\dfrac{3}{12} - \dfrac{1}{2} =$ _____

d $1\dfrac{1}{3} - \dfrac{5}{6} =$ _____

e $3\dfrac{1}{4} - 2\dfrac{1}{2} =$ _____

f $1\dfrac{5}{12} - \dfrac{2}{3} =$ _____

---

3 a Reduce the fractions in the tables below to their simplest form by following these instructions for each fraction:

- Write all the factors of the numerator and of the denominator. (The factors are the numbers that they can be divided by.)
- Circle or underline the **highest common factor** (the highest factor that they can both be divided by).
- Simplify the fraction by dividing the numerator and the denominator by the **highest common factor**.

Fraction: $\dfrac{12}{30}$	Factors of the numerator:	Simplest form
	Factors of the denominator:	$\dfrac{12}{30} =$

b

Fraction: $\dfrac{18}{48}$	Factors of the numerator:	Simplest form
	Factors of the denominator:	$\dfrac{18}{48} =$

c

Fraction: $\dfrac{25}{45}$	Factors of the numerator:	Simplest form
	Factors of the denominator:	$\dfrac{25}{45} =$

---

4 Draw a diagram to show that the two fractions in question 3a are equivalent.

Mastery

1   Use the space provided to prove that $\frac{15}{20}$ is **not** equivalent to $\frac{3}{5}$.

2   **a**   Choose five different fractions and multiply each by $\frac{1}{4}$.

      b   Write a brief statement to describe what happens to the numerator and the denominator when you multiply a fraction by $\frac{1}{4}$.

3   **a**   Choose five different fractions and divide each by $\frac{1}{2}$. Here is an example:

$$\frac{1}{7} \div \frac{1}{2} = \frac{1}{7} \times \frac{2}{1} = \frac{2}{7}$$

      b   Write a brief statement to describe what happens to the numerator and the denominator when you divide a fraction by $\frac{1}{2}$.

OXFORD UNIVERSITY PRESS

Practice

1 Shade the following hundred grids to match the decimals and fractions.

a   0.02   b   0.25   c   0.7   d $\frac{45}{100}$

2 Write *True* or *False* next to each of the following:

Remember:
> means "is greater than" and < means "is less than".

a $0.03 < 0.3$ _____

b $\frac{1}{1000} = 0.001$ _____

c $\frac{99}{1000} = 0.99$ _____   d $0.03 > \frac{3}{1000}$ _____

e $\frac{735}{1000} < 0.735$ _____   f $\frac{3}{4} > 0.075$ _____

g $1.35 = 1\frac{35}{100}$ _____   h $4.350 > 4.35$ _____

3 Colour the hundred grid as follows:

a   Colour 0.2 blue.

b   Colour $\frac{10}{100}$ green.

c   Colour $\frac{50}{1000}$ yellow.

d   Colour $\frac{1}{4}$ red.

e   Colour 0.39 purple.

f   Write the unshaded amount as a fraction and as a decimal. _____

4 Order these from **smallest** to **largest**:

| 0.137 | 0.31 | 1.037 | 1.37 | 0.37 | 0.371 |

_____   _____   _____   _____   _____   _____

Challenge

1 Complete this table. Write in the missing fractions and decimals.

	Fraction	Decimal
a	$\frac{9}{1000}$	
b		0.04
c		0.001
d	$\frac{16}{1000}$	
e		0.25
f	$\frac{750}{1000}$	
g		0.099
h	$\frac{1}{2}$	

2 Change the improper fractions in this table to mixed numbers and then to decimals.

	Improper fraction	Mixed number	Decimal
a	$\frac{5}{4}$		
b	$\frac{24}{10}$		
c	$\frac{175}{100}$		
d	$\frac{250}{100}$		
e	$\frac{475}{100}$		
f	$\frac{2750}{1000}$		

3 Change the following fractions to decimals.

a $\frac{2}{5}$ _____   b $\frac{4}{5}$ _____

c $\frac{5}{20}$ _____   d $\frac{15}{20}$ _____

e $\frac{7}{5}$ _____   f $\frac{25}{20}$ _____

OXFORD UNIVERSITY PRESS

Mastery

1 Change the following fractions to decimals. Round them to a maximum of three decimal places.

a $\frac{5}{8}$ _____

b $\frac{5}{7}$ _____

c $\frac{1}{7}$ _____

d $\frac{7}{8}$ _____

e $\frac{7}{12}$ _____

f $\frac{3}{11}$ _____

2 The final two fractions in the last question had repeating (or recurring) decimals, but you rounded them to three decimal places. When there is one digit that repeats, we can place a dot (or a horizontal line) over the digit to show that it recurs: $0.5\dot{3}$ or $0.5\bar{3}$. If two digits repeat, place a dot (or line) over each digit: $0.1\dot{2}\dot{7}$ or $0.1\bar{2}\bar{7}$

Find the decimal equivalent of:

a $\frac{2}{3}$ _____

b $\frac{1}{6}$ _____

c $\frac{2}{11}$ _____

d $\frac{5}{6}$ _____

e $\frac{7}{11}$ _____

f $\frac{4}{9}$ _____

3 Investigate to find three fractions that have recurring decimals. Label your fractions *a*, *b* and *c*.

4 In some countries a 'dot' separates the thousands from the hundreds. That could make things a bit confusing. Imagine ordering 1.375 kg of sausages and being given one thousand three hundred and seventy-five kilograms instead of 1 kg 375 g!
Do some research to find out some of the countries that use a comma to separate the whole numbers and the decimals.

Practice

1 Calculate the answers to the following:

a
```
    1 • 6  5
+   2 • 5  7
_____
      •
```

b
```
  4  5 • 8
+ 2  5 • 9
_____
     •
```

c
```
  2  5 • 3  9
+ 3  0 • 7  7  5
_____
        •
```

d
```
  7  5 • 8
- 2  7 • 7
_____
     •
```

e
```
  7 • 1  5
- 5 • 3  8
_____
    •
```

f
```
  6  4  4 • 0  7
- 2  6  8 • 7  8
_____
        •
```

g
```
  0 • 2  3
- 0 • 1  7
_____
    •
```

h
```
  0 • 6  9
+ 5 • 8  8
_____
    •
```

i
```
  3  0  0 • 6  2
- 1  6  4 • 8  3
_____
        •
```

2 Use place value strategies to write algorithms for the following:

a 27.3 + 4.5

b 3.45 + 14.2

c 31.08 + 17.5

d 17.36 – 8.55

e 16.2 – 7.57

f 24.3 – 15.7

3

a Building A is 27.34 m high. Building B is 56.2 m high. What is the difference in their heights?

b Three boxes weigh 2.56 kg, 17.065 kg and 9.3 kg. What is their total mass?

OXFORD UNIVERSITY PRESS

Challenge

1 Alice can run twice around some cones on the school athletics track in 19.05 seconds. She does the first lap in 9.47 seconds. How long does the second lap take?

2 A party barrel holds 3.5 L of juice. Jack pours in 2.225 L from one bottle and $\frac{3}{4}$ L from another. How much more can the barrel hold?

3 Henry needs four pieces of wood that are 85 cm long, 1.2 m long, 1.07 m long and half a metre long. He has a 3.9 m length of wood that he can cut them from. How much shorter or longer is it than he needs?

4 How many 135 g servings of rice can be taken from a 1.5 kg bag?

5 By how much is 3.75 kg lighter than 4225 g? Write your answer in grams and in kilograms with decimal notation.

Getting to 10 (Part one)

The answer to each of the equations below is exactly 10. Write one digit in each gap to complete the equations. Try to find at least five different solutions.

0. _____ + 5. _____ 3 + _____ . 27 = 10

0. _____ + 5. _____ 3 + _____ . 27 = 10

0. _____ + 5. _____ 3 + _____ . 27 = 10

0. _____ + 5. _____ 3 + _____ . 27 = 10

0. _____ + 5. _____ 3 + _____ . 27 = 10

Getting to 10 (Part two)

The answer to this puzzle is 10, but you are not given any digits to work with. The rules of the puzzle are:

- One different digit takes the place of a, b, c and d.
- The numbers are read horizontally and vertically as decimals: a.b, c.d, a.c and b.d.
- The sum of the four numbers (a.b + c.d + a.c + b.d) must equal exactly 10.

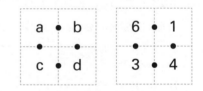

If we try the digits 6, 1, 3 and 4, the sum becomes 6.1 + 3.4 + 6.3 + 1.4. The total is clearly greater than 10, so other digits must be used. Try to find digits that will solve the problem. If you manage to find a solution, try another. There are quite a few ways to succeed. Happy hunting!

OXFORD UNIVERSITY PRESS

Practice

Remember: Put the decimal point in the correct place in the answer box.

1 Multiply the decimals. Look for patterns in the answers.

a
```
      4 • 1
  ×     3
  _____
```

b
```
      0 • 2  6
  ×        9
  _____
```

c
```
      6 • 9
  ×     5
  _____
```

d
```
      6 • 1  7
  ×        2
  _____
```

e
```
      0 • 4  6  9
  ×           5
  _____
```

f
```
      0 • 5  7  6
  ×           6
  _____
```

g
```
      7  6 • 1  3
  ×           6
  _____
```

h
```
      2  9 • 2  1  8
  ×              3
  _____
```

i
```
      1  9 • 7  5  3
  ×              5
  _____
```

2 Divide the decimals. Look for patterns in the answers.

a 4)4 8 . 4

b 3)6 . 9 3

c 5)1 7 1 . 5

d 6)2 7 . 2 4

e 6)3 3 . 9

f 7)4 7 . 3 2

g 8)6 2 . 9 6

h 5)4 4 . 9

i 7)6 3 . 7 6 3

j 4)4 9 . 2 8 4

k 6)1 4 0 5 . 9 2

l 3)1 0 3 . 6 2 9

3 A prize of $4734 is shared by 5 people. How much less than $1000 does each person receive? _____

1 Look for patterns of digits in these multiplications.

a 1212.12 × 7

b 2323.23 × 7

c 34.3434 × 9

d 51.5151 × 6

e 4545.45 × 5

f 565.656 × 9

g 37.0037 × 3

h 37.0037 × 9

i 37.0037 × 6

2 Find the cost of **each individual item** below. Round the answers to an appropriate amount of money.

a Three shirts for $39.99 _____

b Four games for $73.99 _____

c Six chairs for $129.99 _____

d Eight books for $99.99 _____

e Nine balls for $23.99 _____

f Seven balls for $19.95 _____

g Five hats for $47.99 _____

h Six pictures for $199.95 _____

OXFORD UNIVERSITY PRESS

1 Rafferty has $145. He wants to buy some cricket balls. They cost $29.35 each. How many can he afford to buy? Show your working out.

Making one whole

2 The following nine numbers include whole numbers and decimals:

0.1666   0.3333   0.5   0.6666   1   1.5   2   3   6

It is possible to multiply them in sets of three so that the answer is always 1. The challenge is to arrange them on a 3 × 3 grid so that every row, every column and both diagonals have a product of 1.

They could be placed on the grid as shown in the example on the right.

1.5	3	2
1	0.1666	0.5
0.6666	0.3333	6

However, if we multiply the top row, we can easily see that the product is not 1 (1.5 × 3 × 2 = 9), so they must be arranged in a different way. Use the grids below to try to find a solution.
(Hint: If we use the numbers 0.5, 6 and 0.3333, the answer to 0.5 × 6 = 3. And 3 × 0.3333 = 0.9999, which can be appropriately rounded to 1.)

Practice

> Remember: Only use a zero when it is necessary.

Solve these multiplication problems.

1
a $4.7 \times 10 =$ _____

b $4.7 \times 100 =$ _____

c $4.7 \times 1000 =$ _____

2
a $9.3 \times 10 =$ _____

b $9.3 \times 100 =$ _____

c $9.3 \times 1000 =$ _____

3
a $2.39 \times 10 =$ _____

b $2.39 \times 100 =$ _____

c $2.39 \times 1000 =$ _____

4
a $3.001 \times 10 =$ _____

b $3.001 \times 100 =$ _____

c $3.001 \times 1000 =$ _____

> Dividing by **10** moves each digit **one** place smaller. Dividing by **100** moves each digit **two** places smaller and by **1000** moves each digit **three** places smaller.

Solve these division problems:

5
a $234.2 \div 10 =$ _____

b $234.2 \div 100 =$ _____

c $234.2 \div 1000 =$ _____

6
a $423 \div 10 =$ _____

b $423 \div 100 =$ _____

c $423 \div 1000 =$ _____

7
a $7234.8 \div 10 =$ _____

b $7234.8 \div 100 =$ _____

c $7234.8 \div 1000 =$ _____

8
a $260.5 \div 10 =$ _____

b $260.5 \div 100 =$ _____

c $260.5 \div 1000 =$ _____

Solve these multiplication and division problems.

9
a $\$2.35 \times 100 =$ _____

b 0.003 tonnes $\times 1000 =$ _____

c 1.05 m $\times 10 =$ _____

10
a $\$1750 \div 1000 =$ _____

b 34 tonnes $\div 10 =$ _____

c 524.2 km $\div 100 =$ _____

OXFORD UNIVERSITY PRESS

Challenge

1 If you want to calculate 25 × 30, you can split 30 into 3 tens:
Multiply by 10 first: 25 × 10 = 250. **Then multiply by 3**: 250 × 3 = 750.
The same process works for multiplying decimals by powers of 10. For example, if
you want to calculate 2.5 × 30, you can split 30 into 3 tens: **Multiply by 10 first**:
2.5 × 10 = 25. **Then multiply by 3**: 25 × 3 = 75

	× 30	First × 10	Then × 3	Multiplication fact
e.g.	2.5	25	75	2.5 × 30 = 75
a	1.2			
b	2.3			
c	3.1			
d	1.5			
e	0.32			

2 Sometimes it's easier to split the multiple of 10 differently. 2.5 × 30 is the same as finding
2.5 × 3 ten times. **Multiply by 3 first**: 2.5 × 3 = 7.5. **Then multiply by 10**: 7.5 × 10 = 75

	× 30	First × 3	Then × 10	Multiplication fact
e.g.	2.5	7.5	75	2.5 × 30 = 75
a	1.3			
b	2.2			
c	3.2			
d	5.1			
e	0.33			

3 Use your choice of strategy to multiply these decimals by multiples of 10.

a   1.7 × 40 = _____

b   2.5 × 40 = _____

c   3.5 × 50 = _____

d   3.4 × 30 = _____

e   1.6 × 30 = _____

f   12.5 × 20 = _____

g   1.5 × 50 = _____

h   7.2 × 20 = _____

i   1.7 × 30 = _____

j $1.20 × 50 = _____

k $2.25 × 30 = _____

l   2.1 L × 40 = _____

m   8.3 kg × 30 = _____

n   24.8 cm × 60 = _____

4 A factory sends out 40 packages. Each one has a mass of 1.275 kg. What is the total
mass of the packages? _____

Here is a reminder of how you can show that 21 × 15 = 315:

		2	1
×		1	5
	1	0	5	← 21 × 5
+	2	1	0	← 21 × 10
	3	1	5

The process is the same for multiplying decimals by two digits. Here is how you can show that 2.1 × 15 = 31.5:

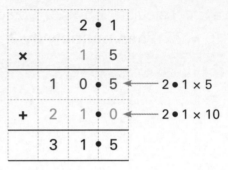

1 Multiply:

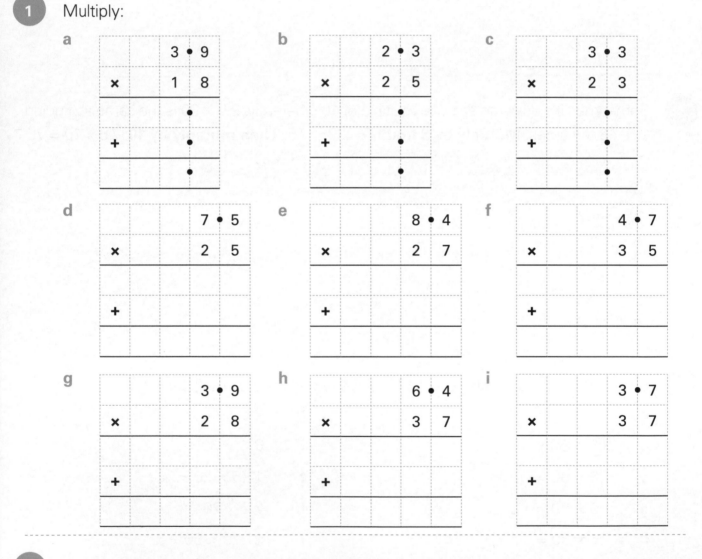

a

		3 • 9
×		1 8
		•
+		•
		•

b

		2 • 3
×		2 5
		•
+		•
		•

c

		3 • 3
×		2 3
		•
+		•
		•

d

		7 • 5
×		2 5
+		

e

		8 • 4
×		2 7
+		

f

		4 • 7
×		3 5
+		

g

		3 • 9
×		2 8
+		

h

		6 • 4
×		3 7
+		

i

		3 • 7
×		3 7
+		

2 Nancy is saving for a car. She saves $125.65 each week for a whole year. The car she wants is $6699. How much more or less than the cost of the car does she have by the end of the year?

OXFORD UNIVERSITY PRESS

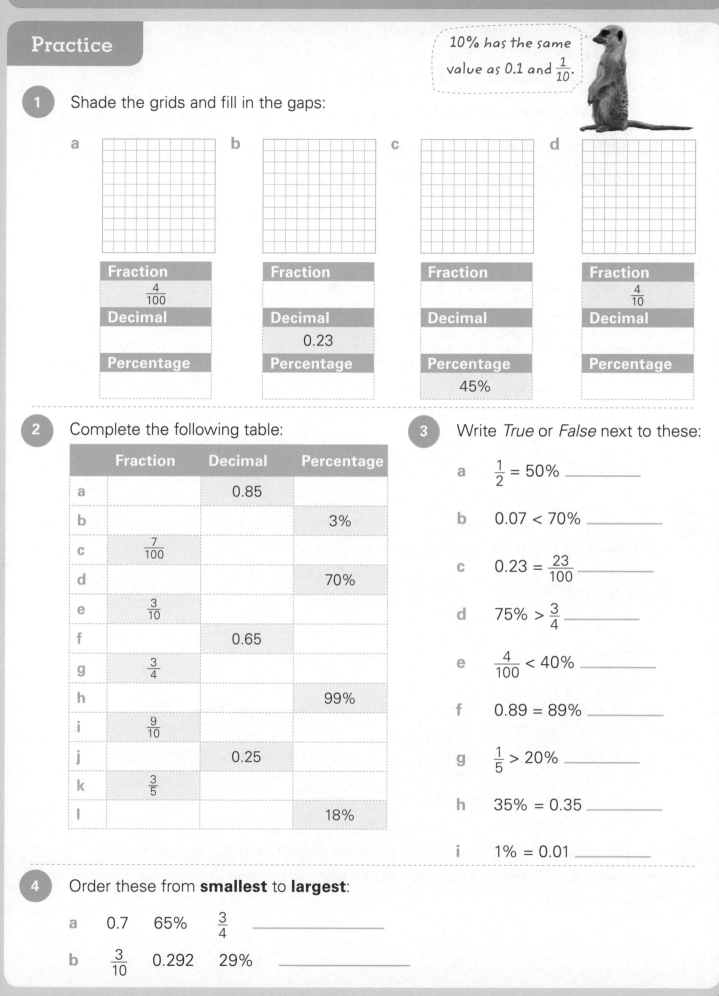

Practice

10% has the same value as 0.1 and $\frac{1}{10}$.

1 Shade the grids and fill in the gaps:

a
Fraction
$\frac{4}{100}$
Decimal
Percentage

b
Fraction
Decimal
0.23
Percentage

c
Fraction
Decimal
Percentage
45%

d
Fraction
$\frac{4}{10}$
Decimal
Percentage

2 Complete the following table:

	Fraction	Decimal	Percentage
a		0.85	
b			3%
c	$\frac{7}{100}$		
d			70%
e	$\frac{3}{10}$		
f		0.65	
g	$\frac{3}{4}$		
h			99%
i	$\frac{9}{10}$		
j		0.25	
k	$\frac{3}{5}$		
l			18%

3 Write *True* or *False* next to these:

a $\frac{1}{2}$ = 50% _____

b 0.07 < 70% _____

c 0.23 = $\frac{23}{100}$ _____

d 75% > $\frac{3}{4}$ _____

e $\frac{4}{100}$ < 40% _____

f 0.89 = 89% _____

g $\frac{1}{5}$ > 20% _____

h 35% = 0.35 _____

i 1% = 0.01 _____

4 Order these from **smallest** to **largest**:

a 0.7 65% $\frac{3}{4}$ _____

b $\frac{3}{10}$ 0.292 29% _____

Challenge

1 Yale conducted a survey of 150 people. Coffee was the preferred hot drink of 112 of those surveyed.

    a Express the fraction, in its lowest form,

        of people who preferred coffee. _____

    b Estimate the decimal and percentage

        of people who preferred coffee. _____

    c Use a calculator to find the exact decimal and percentage. _____

2 Fill in the blanks according to their position on this number line. If an arrow does not point to an exact decimal on the line, estimate its position. The first blank has been filled in for you—it is 5% of the way along the number line.

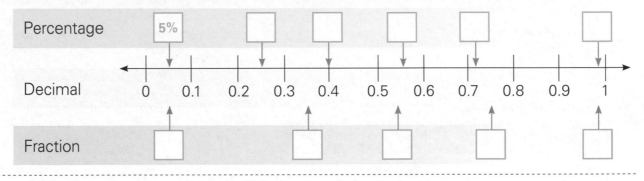

3 The triangle is 25% of the way along the number line. Choose the best estimate for the positions of the following shapes:

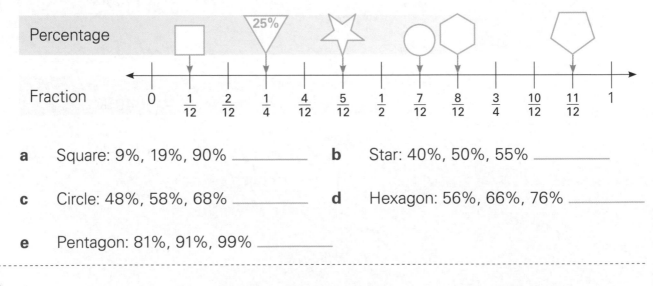

    a Square: 9%, 19%, 90% _____   **b** Star: 40%, 50%, 55% _____

    c Circle: 48%, 58%, 68% _____   **d** Hexagon: 56%, 66%, 76% _____

    e Pentagon: 81%, 91%, 99% _____

4 Find the decimal equivalents of the positions of the five blank shapes on the number line in question 3 (rounded to two decimal places). Convert the decimals to percentages.

OXFORD UNIVERSITY PRESS

The table below shows the size of some of the world's largest countries. It also shows the area that is covered by forest, and the area of inland waterways in each country.

	Land area (to nearest $\frac{1}{2}$ million km²)	Area which is forest (to nearest $\frac{1}{2}$ million km²)	Area which is forest (to nearest 1%)	Area of inland water (to nearest 5000 km²)	Area which is inland water (to nearest 0.1%)
Russia	17 million km²	8 million km²			0.5%
Brazil	8.5 million km²	5 million km²			0.7%
Canada	10 million km²	3 million km²			9%
USA	10 million km²		30%	660 000 km²	
China	9.5 million km²		21%	270 000 km²	
Australia	7.5 million km²		20%	70 000 km²	

1 From the table we can see that almost half of Russia is covered by forest. The actual fraction can be found by dividing 8 million by 17 million. Multiplying the answer by 100 will give the percentage.

   a Fill in the percentage of forest area for Russia, Brazil and Canada. Round to the nearest 1%.

   b Fill in the area of forest in the table for USA, China and Australia. Round to the nearest half million km².

   c Fill in the percentage of inland water area in the table for USA, China and Australia. Round to the nearest 0.1%.

   d Fill in the area of inland water for Russia, Brazil and Canada. Round to the nearest 5000 km².

2 The figures in the table above are all rounded. Research the actual figures for one or two countries. Find out how close the actual figures are to the figures in the table. Comment on how important it is to have true figures when showing information like this.

Practice

1 a   Write the quantity of trucks to motorbikes as a ratio. _____

   b   Rewrite the ratio in its simplest form. _____

2   Use the ratio to work out how many motorbikes there would be if there were:

   a   10 trucks _____   b   100 trucks _____

   c   25 trucks _____

3   Use the ratio to work out how many trucks there would be if there were:

   a   30 motorbikes _____   b   100 motorbikes _____

   c   150 motorbikes _____

4   Imagine 15 cars were added to the trucks and motorbikes.

   a   Write the ratio of trucks to motorbikes to cars in its simplest form.

        _____

   b   How many trucks would there be if there were 45 cars? _____

   c   How many motorbikes would there be if there were 60 cars? _____

5   Colour the whole grid with a ratio of 5 red squares to 2 blue squares to 3 green squares (5:2:3).

OXFORD UNIVERSITY PRESS

Challenge

1 Mr Musto's class had students with red hair, blonde hair, black hair and brown hair in the following ratio:

$$2:5:7:11$$

If the entire school had the same ratio, how many students with each hair colour would there be if the school had the following number of students?

	Number of students	Red hair	Blonde hair	Black hair	Brown hair
a	125				
b	200				
c	300				
d	375				

2 At Tom's party, 12 guests chose lemonade and 8 chose orange juice.

a What is the ratio of guests who had lemonade to total guests in its simplest form? _____

b What is the ratio of guests who had orange juice to guests who had lemonade in its simplest form? _____

c What proportion of guests had lemonade as a percentage? _____

3 At Lucia's party, 26 guests had pizza, 18 had pasta and 6 had salad.

a What is the ratio of guests who had pasta to the total guests in its simplest form? _____

b What is the ratio of guests who had pizza to the total guests in its simplest form? _____

c What is the ratio of guests who had salad to the total guests in its simplest form? _____

d What is the ratio of guests who had pizza to guests who had pasta and salad in its simplest form? _____

e What proportion of guests had salad as a decimal? _____

1   Olivia had a box of 144 cookies. Some were chocolate, some were vanilla and some were strawberry. What might the ratio of each flavour have been and how many of each cookie might there have been? Show at least five solutions.

2   Quan had 75 marbles. Some were red and some were green. What might the proportion of each colour have been as a fraction, percentage and decimal? Show at least three solutions.

3   Write the ratios for each of your solutions from question 2.

OXFORD UNIVERSITY PRESS

Geometric and number patterns

Practice

1 Write a rule for this pattern of sticks.

_____

2 **a** Look at the pattern above and complete the following table by working out how many sticks you would need to make the number of pentagons specified.

Number of pentagons	1	2	3	4	5	6	7	8	9	10
Number of sticks	5									

  b How many sticks would be needed for a pattern of 20 pentagons? _____

3 Look at the way this pattern has been made.

A rule for making the pattern could be:

Use five sticks for the first pentagon and then four for every other pentagon.

Which of these rules could also be used to describe the way pattern is made? (Circle one.)

- Use five sticks for every pentagon.

- Use five sticks for the first pentagon, four for the second pentagon, three for the third pentagon, two for the fourth pentagon and one less for every other pentagon.

- Start with one stick and then use four sticks for every pentagon.

4 Follow this rule for the patterns of pentagons and then write a number sentence to show how many sticks are needed for each pattern.

Pattern	Rule	How many sticks are needed?
a	Use five sticks for the first pentagon and then four for every other pentagon.	$5 + 4 \times \boxed{} - \boxed{}$
b		$5 + \boxed{} = \boxed{}$ $5 + \boxed{} \times \boxed{} = \boxed{}$

5 Write a rule for this pattern.

_____

Challenge

1 This flow chart shows a rule of divisibility by 9. It can be used for testing whether any number is divisible by 9.

Use the flow chart to decide whether or not these numbers are divisible by 9. Answer *Yes* or *No*. Prove that the rule works for any number by doing a division that shows whether or not there is a remainder.

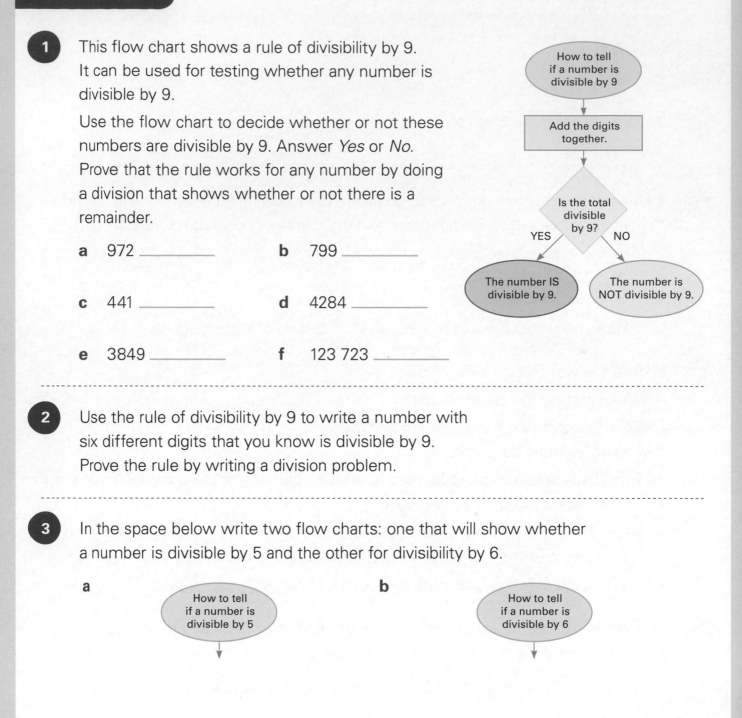

a 972 _____

b 799 _____

c 441 _____

d 4284 _____

e 3849 _____

f 123 723 _____

2 Use the rule of divisibility by 9 to write a number with six different digits that you know is divisible by 9. Prove the rule by writing a division problem.

3 In the space below write two flow charts: one that will show whether a number is divisible by 5 and the other for divisibility by 6.

a

How to tell if a number is divisible by 5

b

How to tell if a number is divisible by 6

OXFORD UNIVERSITY PRESS

1   You will probably recognise the numbers 2, 3, 5, 7 and 9 as the single-digit prime numbers. But what about two-digit numbers? How can you work out whether or not they are prime numbers? First, you must ask: *Is the number divisible by 2?*

Design a flow chart that will serve as a test to show whether any two-digit number is a prime number. It has been started for you below.

A flow chart needs to be neat and easily understood by anyone who reads it.

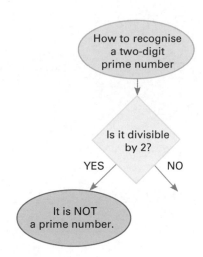

How to recognise a two-digit prime number

Is it divisible by 2?

YES   NO

It is NOT a prime number.

2   Use your flow chart to make a list of as many two-digit prime numbers as you can.

Order of operations

Practice

1 Use the order of operations to complete the following:

a $5 \times 2 + 4 =$ _____

b $3 + 8 \times 2 =$ _____

c $24 \div 4 - 3 =$ _____

d $(6 - 3) + (10 - 3) =$ _____

e $2 + 54 \div 9 =$ _____

f $36 \div (8 - 4) =$ _____

g $25 - 12 \div 3 =$ _____

h $(35 - 7) \div 2 =$ _____

i $\frac{1}{2}$ of $9 + 3 =$ _____

j $\frac{1}{2}$ of $(9 + 3) =$ _____

k $3 \times 3^2 =$ _____

l $(3 \times 3)^2 =$ _____

m $(4 + 5)^2 =$ _____

n $4 + 5^2 =$ _____

o $\frac{1}{4}$ of $12 - 2 =$ _____

p $\frac{1}{4}$ of $(12 - 2) =$ _____

2 Use equations to split the number you are multiplying by.

	Problem		Split the problem to make it simpler		Rewrite the problem		Answer
a	42×3	=	$(40 \times 3) + (2 \times 3)$	=	$120 + 6$	=	
b	14×6	=		=		=	
c	53×5	=		=		=	
d	23×9	=		=		=	

3 Changing the order of operations.

	Problem		Change the order to make it simpler		Rewrite the problem		Answer
a	$25 \times 11 \times 4$	=	$25 \times 4 \times 11$	=		=	
b	$50 \times 12 \times 4$	=		=		=	
c	$8 \times 15 \times 4$	=		=		=	
d	$5 \times 19 \times 2$	=		=		=	

OXFORD UNIVERSITY PRESS

Challenge

1 Rewrite each equation in the table below using "opposites" to find the value of ◆.

	Problem	Use opposites	Write the value of ◆	Check by writing the equation
a	◆ × 7 = 63	◆ = 63 ÷ 7		_____ × 7 = 63
b	◆ − $2\frac{1}{2}$ = 6			
c	$\frac{1}{4}$ of ◆ = 15			
d	◆ × 100 = 225			
e	◆ ÷ 100 = 1.6			
f	◆ ÷ 8 = $\frac{1}{8}$			
g	◆ × 1000 = 7225			

2 Substitution: Put a number in place of the ◆ symbol to see if the equation balances. Circle or highlight your choice for the substitute number. Write a number sentence to check.

	Problem	Possible substitutes for ◆				Check
a	◆ × (2 + 3) = 50	8	9	10	11	
b	72 ÷ ◆ − 5 = 3	9	10	11	12	
c	$1\frac{1}{2}$ × ◆ + 6 = $10\frac{1}{2}$	2	3	4	5	
d	25 ÷ ◆ + 2.5 = 5	5	10	15	20	
e	25 × (10 − ◆) = 150	3	4	5	6	
f	◆ × 2 = 3 + 5^2	12	13	14	15	
g	(5 + ◆) × 10 = 25 × 3	1.5	2	2.5	3	

3 Solve the following by substitution:

a ◆ × 4 = 10^2 _____

b (6 ÷ 4) × ◆ = 3 _____

c 4 + 2 × ◆ = 10 _____

d $\frac{1}{2}$ of 6 × ◆ = 15 _____

Problem-solving

When solving problems, we usually use the technique that is quickest. One method is to take out all the unnecessary information and write an equation.

1 Two classes are going out for sport and need hoops. The sport teacher already knows that they will need 150 hoops, but she gives two students the following problem on a piece of paper:

*There are **27** students in your class and **23** in the other class. We need **three hoops** for **each child**. Please fetch the correct number of hoops.*

They write an equation using just the important information.

The number sentence they write is **27 + 23 × 3**.

However, unless they use the order of operations correctly, they will not get the correct number. Solve both equations, and circle the correct one.

$27 + 23 \times 3 =$ _____ $(27 + 23) \times 3 =$ _____

2 Imagine that the story above changes slightly. The teacher also needs one hoop for herself. Rewrite the equation showing the total number of hoops that are needed.

_____

3 Write equations that will solve the following problems and use them to find the correct answers.

a   Henry has $4 and his aunty gives him another $2. His mother says she will double the amount. The same day Henry gets three birthday cards, each with a $5 note inside. The next day he gets five cards: one has $2, two have $5 notes, and two have $10 notes. How much does he have in total?

b   Finlay has a new novel. It has 96 pages. During the first two days he reads five pages a day. The next day, he reads a quarter of the book. He then reads 10 pages a day for two days. How many pages are there left to read by that time?

Practice

1 Convert between the units of length.

a

cm	mm
	20 mm
3.5 cm	
	47 mm
12.3 cm	
	150 mm

b

m	cm
4 m	
	250 cm
2.32 m	
	435 cm
7.5 m	

c

km	m
	3000 m
14 km	
	4750 m
3.9 km	
	1800 m

2 Match **two** lengths to each object in the table below.

2.7 m   2.7 km

75 mm   27 mm   2700 mm

2700 m   127 cm

2.7 cm   1.27 m   7.5 cm

	Object	1st unit	2nd unit
a	The length of a paper clip		
b	The height of a young student		
c	The length of a finger		
d	The height of a ceiling		
e	The length of a street		

3 a List three different measuring tools that you can use to measure length.

_____ _____ _____

b Which one of these tools would you use to measure the width of this page? Give a reason for your choice.

Challenge

1 Circle any of the following that do **not** describe the length of a 3.05 m room.

305 cm   3 m 5 cm   3 m 50 cm   3050 mm   350 cm

2 Line B is 8.5 cm long.

a Estimate (DO NOT MEASURE) the lengths of the other two lines.

Line A _____ _____

Line B _____

Line C _____ _____

b Write the lengths of Lines A and C to the nearest millimetre.

Line A: _____   Line C: _____

3 **a** The perimeter of the triangle is 9 cm. What is the smallest number of sides that you would need to measure to calculate the perimeter? _____

b What is the length of each side of the triangle? _____

4 Find the perimeter of each shape below. Write down the number of sides that you measured to calculate the perimeter of each shape.

a

Perimeter = _____

Number of sides
I measured: _____

b

Perimeter = _____

Number of sides
I measured: _____

5 Draw a rectangle with a perimeter of exactly 34.8 cm. Use the line below as one of the sides of your rectangle. Write the lengths on one long and one short side of the rectangle.

OXFORD UNIVERSITY PRESS

Mastery

1 This is the Peel P50—the smallest production car ever made. Production started and ended in the 1960s. They are very rare, and can sell at auction for more than $100 000!

The Peel P50 is 1.34 m long, 1.2 m high and only 0.99 m wide.

    a   Write your height in cm. _____

    b   What is the difference between your height and the height of the P50? _____

    c   Write the width of the P50 using a different unit of length. _____

    d   What is something in your room that is about the same as the length of the P50 car? _____

2 In 2012 an even smaller car was built by Austin Coulson from Phoenix, Arizona, USA. It is allowed on the roads, but it cannot be called a production car because only one was ever made. The car is 7.53 cm shorter and 56.5 cm lower than the P50. (It has no roof, of course!) It is also an amazing 33.59 cm narrower than the tiny Peel P50.

    a   What are the dimensions of Austin Coulson's car? _____

    b   With your teacher's permission, search online for "Smallest car, Austin Coulson, 2012" to have a look at the car.

    c   What is something in a home or a classroom that is about the size of the smallest car? _____

3 The smallest book in your classroom would probably not be as small as the one in this photo!

The smallest comic book was made in the 1990s by Martin Lodewijk. It is only 37 mm high. Its width is 1.12 cm less than its height. Write the width of the comic book in cm and in mm. _____

4 3D printers can make amazingly small objects —so small that they cannot even be measured in millimetres. They are measured in **micrometres**. The smallest replica guitar in the world was made using a 3D printer at a university in New York. It's a very tiny version of the guitar in this photo, but it is actually smaller than one of the full stops on this page!

The tiny model guitar is only 10 micrometres long! Find out what you can about micrometres so that you can appreciate how amazing the replica guitar is.

Practice

1 Measure each rectangle below to calculate its area.

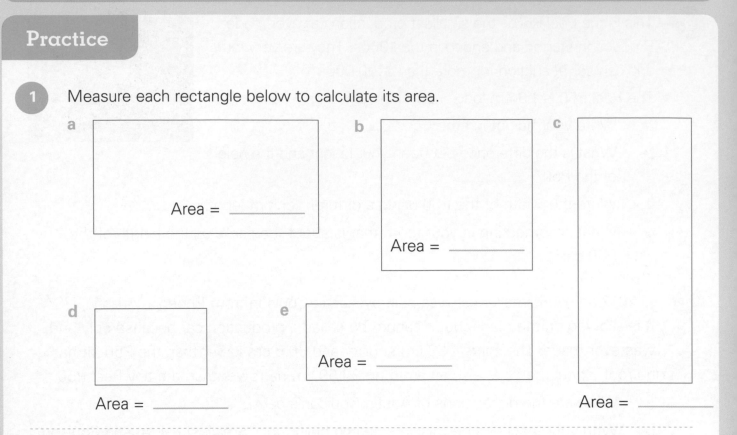

a

Area = _____

b

Area = _____

c

Area = _____

d

Area = _____

e

Area = _____

2 This is a floor plan of a housing unit. Using a scale of 1 cm: 1 m, write the area of each room.

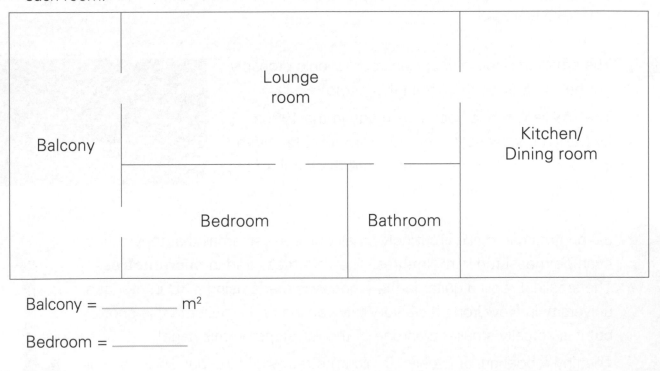

Balcony = _____ m²

Bedroom = _____

Bathroom = _____

Lounge room = _____

Kitchen/Dining room = _____

OXFORD UNIVERSITY PRESS

Challenge

1 Measure and then calculate the area of each shape.

a

Area = _____

b

Area = _____

c

Area = _____

d

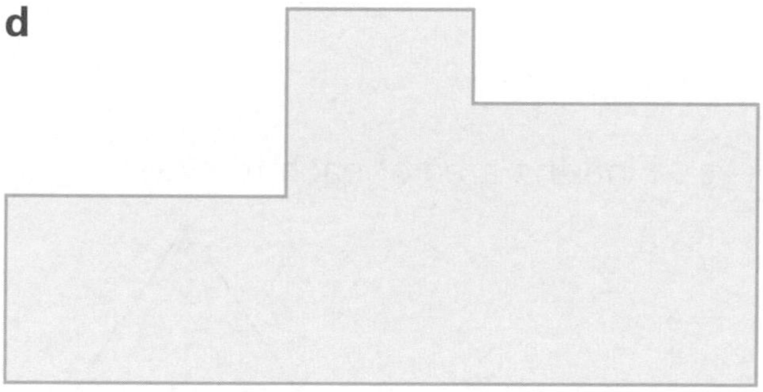

Area = _____

e

Area = _____

2 Draw two different shapes that are not rectangles. Each shape must have an area of 24 cm².

1 Find the area of each right-angled triangle.

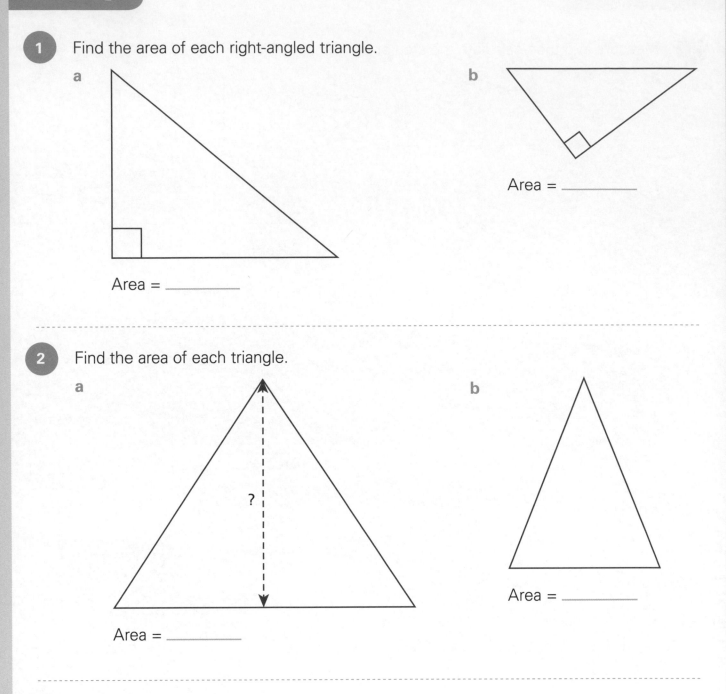

a

Area = _____

b

Area = _____

2 Find the area of each triangle.

a

?

Area = _____

b

Area = _____

3 Design a pen for a dog to run around in. You have 36 m of chicken wire and plenty of stakes.

Start with a rectangular enclosure. A pen 17 m long by 1 m wide would use all the wire, but would other dimensions provide a bigger area?

Draft some plans for the dog pen. Find one that provides the biggest area whilst using all 36 m of wire. Using a suitable scale, draw it on a separate piece of paper. (Grid paper might be useful.)

You could also design a non-rectangular shape so that the dog would have places to hide, sleep, feed and play. On your final copy, show the position of the kennel, the play area etc.

OXFORD UNIVERSITY PRESS

Practice

1 **a** How many centimetre cubes would be needed to make this model? _____

    b What is its volume? _____

1 cm
3 cm
6 cm

2 Find the volume of each of the following objects:

a

5 cm
2 cm
8 cm

Volume: _____ cm³

b

3 cm
3 cm
7 cm

Volume: _____ cm³

c

3 cm
4 cm
7 cm

Volume: _____ cm³

3 Complete the following tables to convert between the different units of volume and capacity.

a

Kilolitres	Litres
5 kL	
2.5 kL	
	4250 L
	3750 L

b

Litres	Millilitres
4 L	
	3500 mL
2.25 L	
	9750 mL

c

Volume	Capacity
100 cm³	mL
	500 mL
175 cm³	
	2 L

4 Put these capacities in order from smallest to largest:

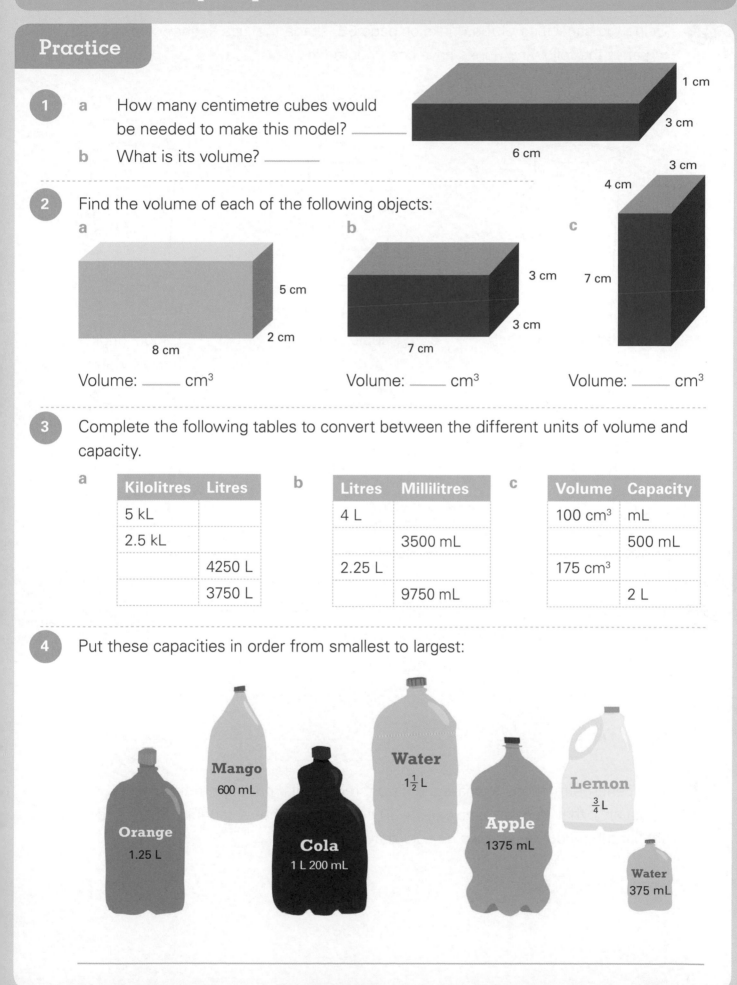

Mango 600 mL

Cola 1 L 200 mL

Water 1½ L

Apple 1375 mL

Lemon ¾ L

Orange 1.25 L

Water 375 mL

Challenge

1 Using the capacities of the drinks on page 68, shade the jugs to show the level in the jug after the following drinks have been poured in:

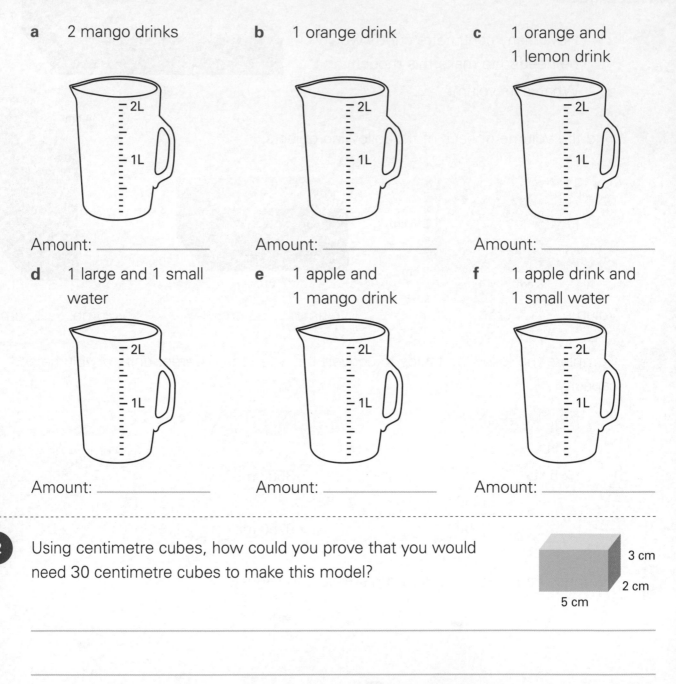

a 2 mango drinks

Amount: _____

b 1 orange drink

Amount: _____

c 1 orange and 1 lemon drink

Amount: _____

d 1 large and 1 small water

Amount: _____

e 1 apple and 1 mango drink

Amount: _____

f 1 apple drink and 1 small water

Amount: _____

2 Using centimetre cubes, how could you prove that you would need 30 centimetre cubes to make this model?

3 cm
2 cm
5 cm

_____

_____

3 Use centimetre cubes to make a different rectangular prism from the one in question 2 that also has a volume of 30 cm³. Sketch the rectangular prism and write the length, width and height on your drawing.

OXFORD UNIVERSITY PRESS

1 Very large quantities of water are measured in megalitres (ML) and gigalitres (GL). Conduct some research to find some information about these units of capacity, such as how they compare with other units of capacity, when they are used in real life, and how important it is to measure accurately when using them in real-life situations.

2 Taking as much care as possible, sketch as many **different** rectangular prisms as you can that have a volume of 24 cm³. Write the length, width and height on each model.

Mass

Practice

1 Complete the tables to convert between the different units of mass.

a

Tonnes	Kilograms
	2000 kg
3.5 t	
	4250 kg
5.175 t	
	975 kg

b

Kilograms	Grams
5 kg	
	3500 g
0.75 kg	
	450 g
3.07 kg	

c

Grams	Milligrams
3 g	
	2250 mg
1.5 mg	
	735 mg
	1 mg

2 Record the mass of each box below, taking note of the increments on the scales.

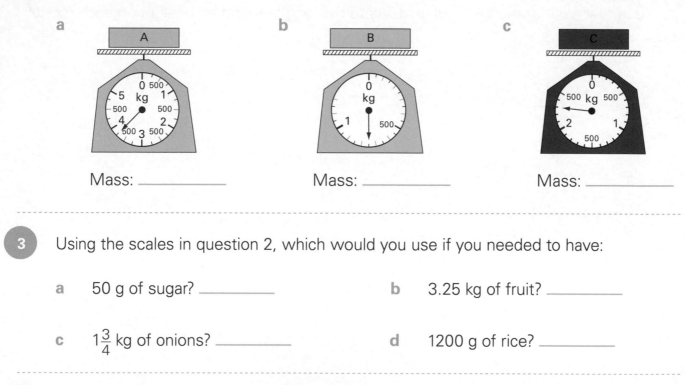

a

Mass: _____

b

Mass: _____

c

Mass: _____

3 Using the scales in question 2, which would you use if you needed to have:

a 50 g of sugar? _____

b 3.25 kg of fruit? _____

c $1\frac{3}{4}$ kg of onions? _____

d 1200 g of rice? _____

4 Draw a pointer on the scale to show that the box has a mass of 4.4 kg.

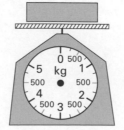

OXFORD UNIVERSITY PRESS

Challenge

1 Estimate the mass of each box and write it in as many different ways as you can.

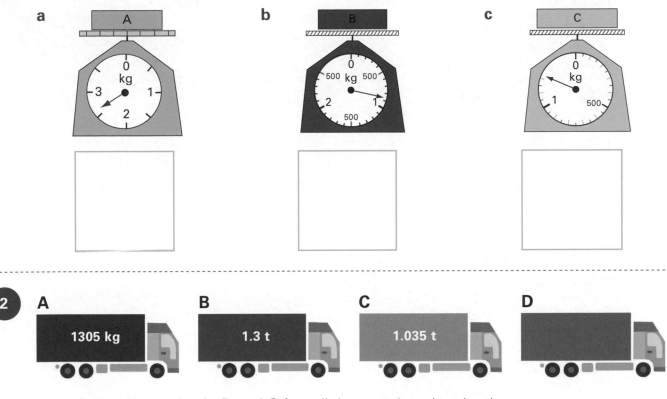

a A

b B

c C

2

A [1305 kg] B [1.3 t] C [1.035 t] D

a Order the trucks A, B and C from lightest to heaviest load. _____

b Truck D has a load that is greater than the load of Truck B but less than the load of Truck A. Write a possible mass for Truck D's load in tonnes with a decimal; in tonnes and kilograms; and in kilograms. _____

_____

c What is the difference between the mass of the loads on trucks B and C?

_____

---

3 You will need a set of weighing scales for this task. Find three items in your classroom (not weights) that you think will have a total mass of 1 kilogram.

a Write your estimate for the mass of each in the table, making sure that the total is 1 kg.

b Weigh each object and record its mass.

c How close to 1 kg was your estimate? _____

Object	My estimate	Actual mass
Total		

1 Match each truck to its load.

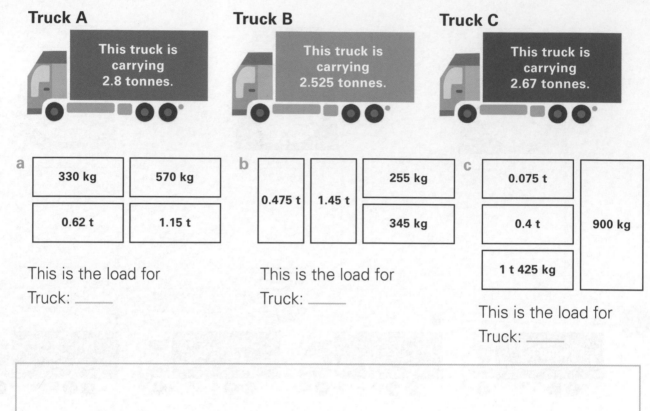

Truck A

This truck is carrying 2.8 tonnes.

Truck B

This truck is carrying 2.525 tonnes.

Truck C

This truck is carrying 2.67 tonnes.

a

| 330 kg | 570 kg |
| 0.62 t | 1.15 t |

This is the load for
Truck: _____

b

| 0.475 t | 1.45 t | 255 kg |
| | | 345 kg |

This is the load for
Truck: _____

c

0.075 t	
0.4 t	900 kg
1 t 425 kg	

This is the load for
Truck: _____

2 a Write the total mass of the 12 large apples in kilograms with a decimal; in kilograms and grams; and in grams.

_____

b What is the mean average mass of one apple (to the nearest gram)? _____

c No two apples have exactly the same mass. Write a possible mass for each of the 12 apples.

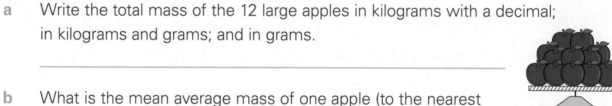

_____

_____

OXFORD UNIVERSITY PRESS

Practice

1 Fill in the missing times.

a

am/pm time   4:34 pm

24-hour time   _____

b During the night

am/pm time   _____

24-hour time   _____

c

am/pm time   _____

24-hour time   13:13

d In the morning

am/pm time   _____

24-hour time   _____

e

am/pm time   11:47 am

24-hour time   _____

f

am/pm time   _____

24-hour time   00:03

The History of Gold Rush Town!

2 Use the information to complete the timeline. Fill in the years and put the correct letters in the boxes. Draw an arrow from each box to the correct place on the timeline.

- 1853: Mining licence fees reduced to 10 shillings—**L**
- 1861: 50 000 prospectors leave—**S**
- 1851: Gold is discovered! Mining licence fee is 30 shillings—**G**
- 1855: 10 000 new miners arrive—**R**
- 1858: Gold is discovered in another town—**U**
- 1854: 28 people are killed in a mining accident—**D**
- 1852: 100 000 prospectors arrive from overseas—**O**
- 1864: Gold runs out in Gold Rush Town—**H**

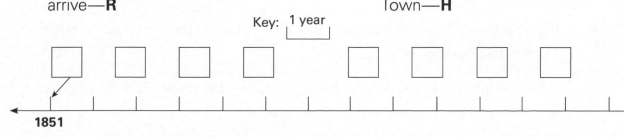

Key: 1 year

1851

Challenge

1 Use the information in this train timetable to complete the tasks below.

a How long does the first train take to get from Springwood to Penrith?

b At what time would you leave Blaxland to reach Smith Plains by quarter to seven?

Station	Time							
Springwood	05:13	05:43	06:13	06:31	06:43	07:01	07:13	07:31
Valley Heights	05:16	05:46	06:16	...	06:46	...	07:16	...
Warrenwood	05:20	05:50	06:20	...	06:50	...	07:20	...
Blaxland	05:24	05:54	06:24	06:40	06:54	07:10	07:24	07:40
Glenbrook	05:29	05:59	06:29	06:45	06:59	07:15	07:29	07:45
Lapstone	05:33	06:03	06:33	...	07:03	...	07:33	...
Smith Plains	05:40	06:10	06:40	06:55	07:10	07:25	07:40	07:55
Penrith	05:44	06:14	06:44	06:59	07:14	07:29	07:44	07:59

c How long is the journey from Springwood to Blaxland if you catch the train that leaves around half past six? _____

d Imagine you want to be at a friend's house in Glenbrook at quarter to eight in the morning. It is a 15-minute walk from Glenbrook station to get there. Which train would you catch from Valley Heights? _____

e Which train would you take from Springwood to get to Lapstone by 7 am?

_____

2 This timetable shows some of the return times from Penrith to Springwood. Use the information from both timetables to complete the tasks below.

a At what time does the 13:10 train from Penrith arrive at Springwood? Answer in am/pm time.

_____

Station	Time							
Penrith	11:10	12:10	13:10	14:10	15:10	16:10	17:10	18:39
Smith Plains	11:13	12:13	13:13	14:13	15:13	16:13	17:13	
Lapstone	11:19	12:19	13:19	14:19	15:19	16:19	17:19	
Glenbrook	11:24	12:24	13:24	14:24	15:24	16:24	17:24	
Blaxland	11:29	12:29	13:29	14:29	15:29	16:29	17:29	
Warrenwood	11:32	12:32	13:32	14:32	15:32	16:32	17:32	
Valley Heights	11:36	12:36	13:36	14:36	15:36	16:36	17:36	
Springwood	11:40	12:40	13:40	14:40	15:40	16:40	17:40	19:02

b If you took the train at about 7 am from Springwood to Penrith, how long would you spend there if you caught the return train at 10 minutes past noon?

_____

c What is the difference between the shortest and longest times to get from Penrith to Springwood? _____

d What do you notice about the longest journey between Penrith and Springwood and the longest journey in the other direction? _____

_____

OXFORD UNIVERSITY PRESS

The first people on the moon

1 In 1969, the Apollo 11 space flight landed the first people on the Moon. All of the clocks used during the mission were set to the time on the east coast of America.

a The list below contains some information from the Apollo 11 mission. The table contains a series of times and dates. Match each event in the list to the corresponding date and time by working out the order in which you think the events must have occurred. Write the letter from the table that corresponds to each event in the space provided.

- Neil Armstrong and Buzz Aldrin's lunar module touched down on the Moon with just a few seconds of fuel left. _____

- Armstrong and Aldrin were relieved when their little spacecraft left the Moon's surface. _____

- The three astronauts splashed down in the Pacific Ocean. _____

- Apollo 11 blasted off from Earth on a four-day journey to the Moon. _____

- Neil Armstrong took a step on the Moon and called it a "giant leap for mankind". _____

- The lunar module separated from the main spacecraft and headed for the Moon's surface. _____

9:37 am 16 July	1:47 pm 20 July	4:18 pm 20 July	10:56 pm 20 July	1:54 pm 21 July	12:55 pm 24 July
A	B	C	D	E	F

b Write the event letters from the table in question 1a on the timeline and, as accurately as possible, draw an arrow to show the date and time of each event. The taller marks represent midnight.

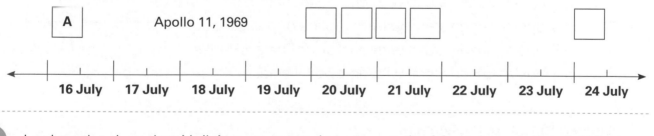

2 Look at the time that Neil Armstrong took a step on the Moon in question 1. What time would it have been in:

a Brisbane, Australia? _____

b London, England? _____

c Moscow, Russia? _____

d Rio de Janeiro, Brazil? _____

Practice

1   Write the name of the shapes below. Draw horizontal stripes on the regular shapes and vertical stripes on the irregular shapes.

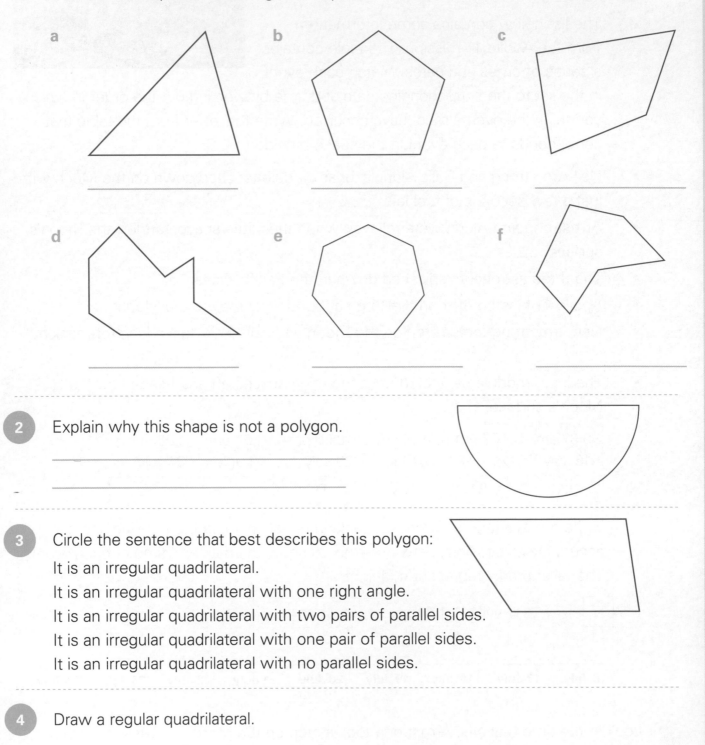

a

b

c

d

e

f

2   Explain why this shape is not a polygon.

_____

_____

3   Circle the sentence that best describes this polygon:

It is an irregular quadrilateral.

It is an irregular quadrilateral with one right angle.

It is an irregular quadrilateral with two pairs of parallel sides.

It is an irregular quadrilateral with one pair of parallel sides.

It is an irregular quadrilateral with no parallel sides.

4   Draw a regular quadrilateral.

Challenge

1 Write the name of each type of triangle and write about the properties that helped you to identify it. You may need a ruler and/or a protractor to help you identify them.

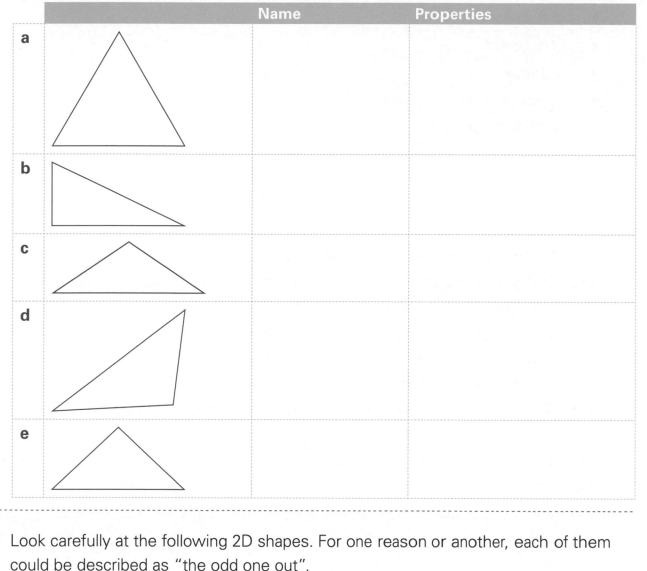

		Name	Properties
a			
b			
c			
d			
e			

2 Look carefully at the following 2D shapes. For one reason or another, each of them could be described as "the odd one out".

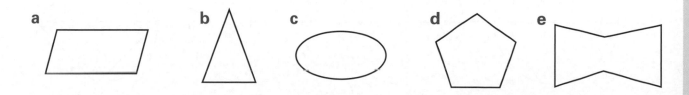

Choose one (or more!) of the shapes and explain why you think it is "the odd one out". _____

3 Draw the shape described below using a pencil and ruler.

- It has four sides.
- Two of the angles are obtuse and two are acute.
- It has two pairs of parallel sides.
- The opposite sides are the same length.
- The base is 9 cm long.

1 In the space below, draw a circle with a radius of 4 centimetres. Label the following:

- the circumference
- a diameter
- a quadrant
- a sector
- a radius
- a semi-circle.

Drawing a perfect square with a pair of compasses

2 It may seem hard to believe, but it is possible to draw a perfect square from a circle! (You will need some plain paper.) Here's how:

Step 1: Draw a circle of any size. Draw a diameter and put two small dots (**A** and **B**) on the diameter. (See Step 1 diagram.)

Step 2: Set your compass at more than half the distance between **A** and **B**. Put the point of the compass at dot **A** and draw arcs above and below the diameter. Repeat from dot **B**. (See Step 2 diagram.)

Step 3: Use a ruler to draw a second diameter that passes through the places where the arcs cross. (See Step 3 diagram.)

Step 4: Join the ends of the two diameters to make a perfect square. (See Step 4 diagram.)

You might need a few practice attempts before drawing your final version.

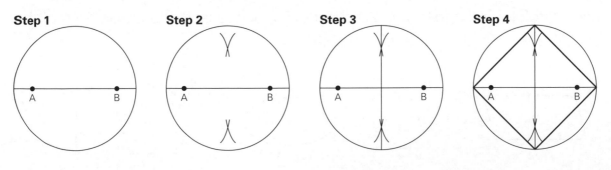

3 It is possible to draw other polygons with the aid of a pair of compasses. You could start, with your teacher's permission, by searching online for "Drawing a hexagon with a pair of compasses".

3D shapes

1 Name these 3D shapes. Say whether each is a *prism*, a *pyramid* or *other 3D shape*.

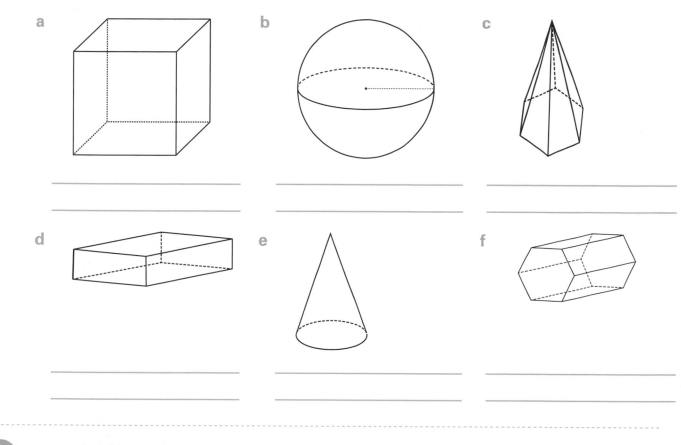

a

b

c

_____

_____

d

e

f

_____

2 Fill in the gaps in the table.

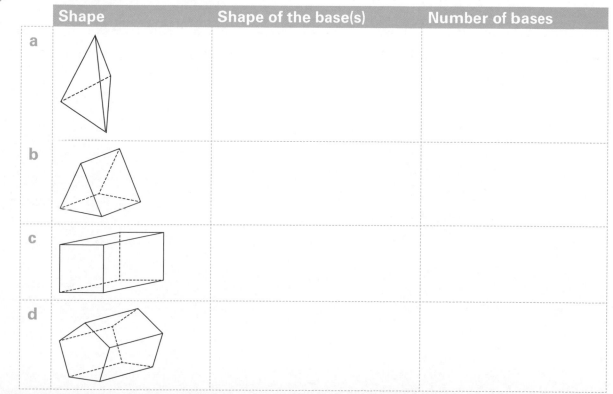

	Shape	Shape of the base(s)	Number of bases
a			
b			
c			
d			

Challenge

1 Copy each shape onto the grid and then make a freehand sketch in the space to the right.

2 The shapes below are not drawn on a grid. Decide on a suitable size and then use the isometric grid to sketch at least two copies of each shape. You may wish to use spare isometric dot paper for further practice.

Cross-sections of 3D shapes

1 Design your own money box using at least two 3D shapes joined together. Draw a picture of your design and write a few sentences describing the shapes you used and why you chose them.

2 The arrows in the drawings below show the direction of the cut for each cross-section. In this question the cuts are all parallel to the base. Identify and draw the cross-section of each 3D shape.

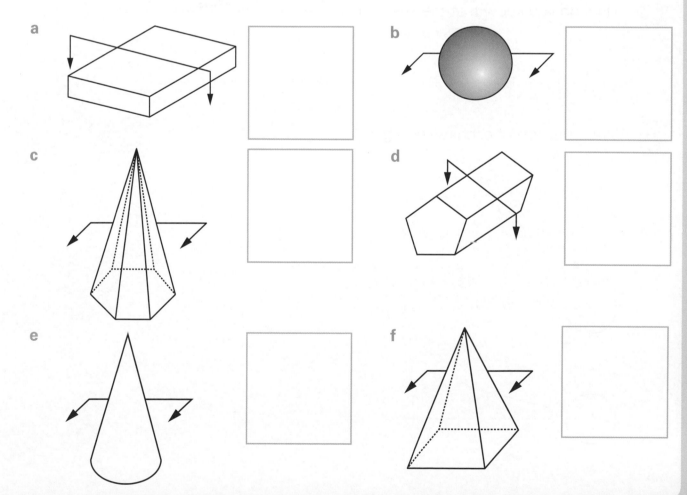

a

b

c

d

e

f

Practice

1 Write the type of each angle. Write the size of the angle inside the arc.

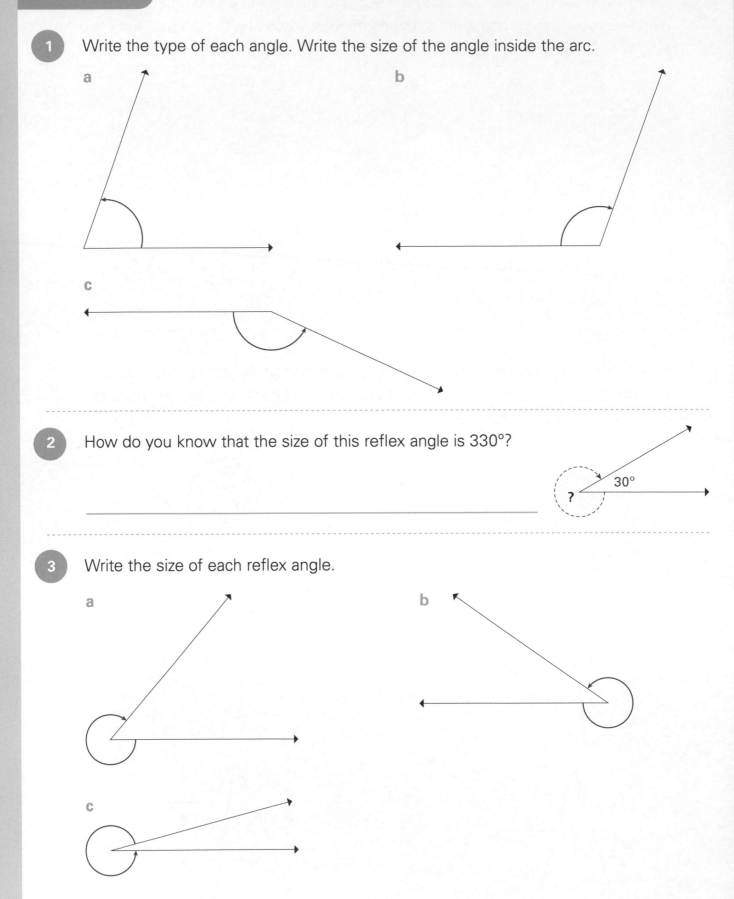

a

b

c

2 How do you know that the size of this reflex angle is 330°?

_____

_____

?   30°

3 Write the size of each reflex angle.

a

b

c

OXFORD UNIVERSITY PRESS

1 Measure the following angles to the nearest degree.

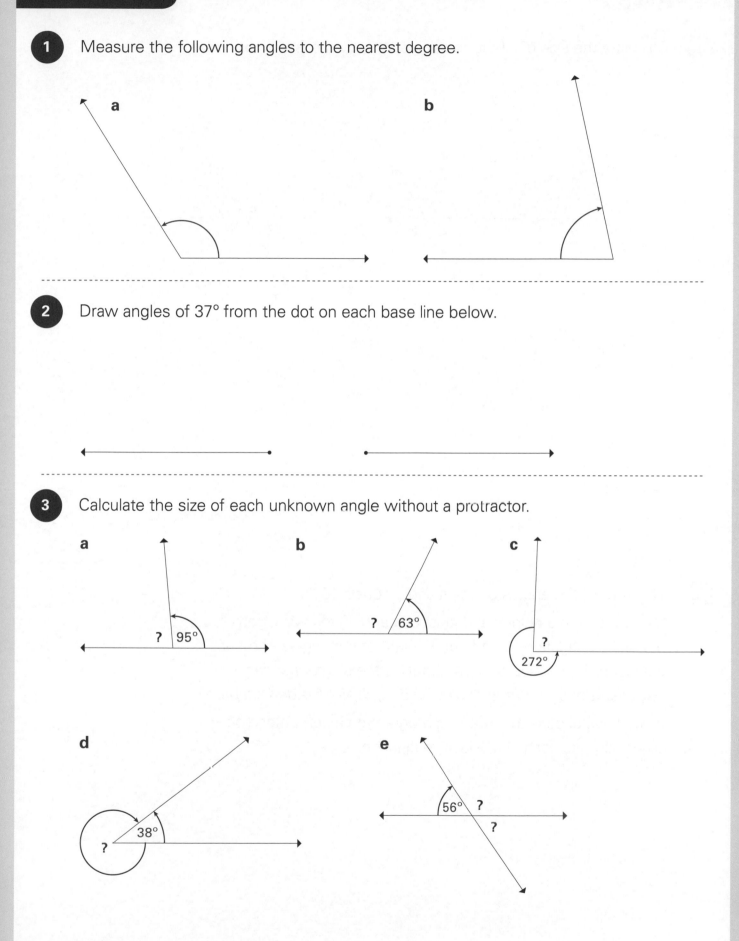

a

b

2 Draw angles of 37° from the dot on each base line below.

3 Calculate the size of each unknown angle without a protractor.

a

? | 95°

b

? | 63°

c

? | 272°

d

38° | ?

e

56° | ? | ?

1   Calculate the size of all the missing angles by measuring **only** the angle marked in red.

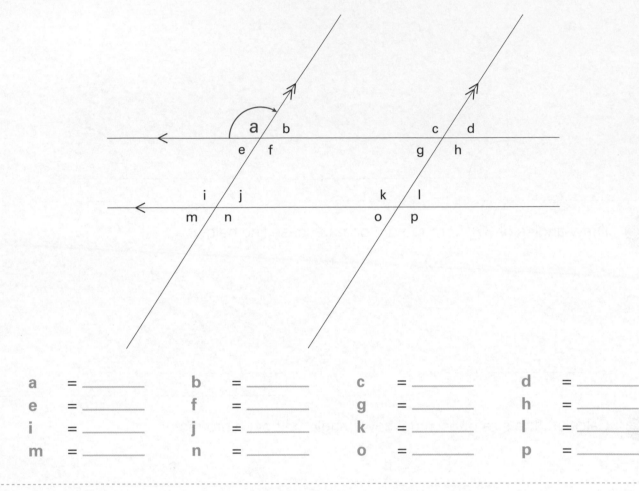

a = _____   b = _____   c = _____   d = _____

e = _____   f = _____   g = _____   h = _____

i = _____   j = _____   k = _____   l = _____

m = _____   n = _____   o = _____   p = _____

2   The sum of the angles of any triangle equals 180°.

Using the space below, and practice paper if you wish, carry out an investigation to find out the sum of the angles of other polygons. Something to think about: *Is there a connection between the sum of the angles on a 3-, 4-, 5-, 6-sided shape?*

If you have access to a drawing program on a computer, you could use it to carry out your investigation.

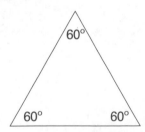

OXFORD UNIVERSITY PRESS

Practice

1 Follow the instructions to make patterns.

    a Reflect this triangle.

    b Now, translate the triangle.

    c Now, rotate the triangle a quarter of a turn clockwise.

2 What method of transformation has been used for the following pictures?

    a

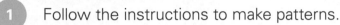

                                                                  _____

    b

                                                                  _____

    c

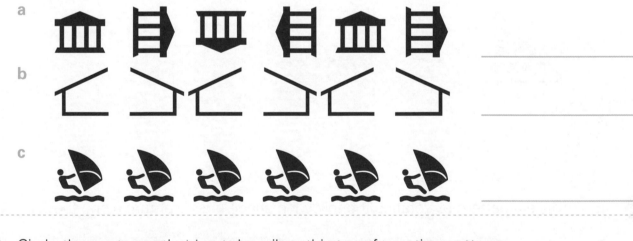

                                                                  _____

3 Circle the sentence that best describes this transformation pattern.

- The shape could have been translated or reflected.
- The shape could have been reflected or rotated.
- The shape could have been translated, reflected or rotated.
- The shape could have been rotated or translated.

1   Make a pattern by reflecting this shape horizontally and vertically.

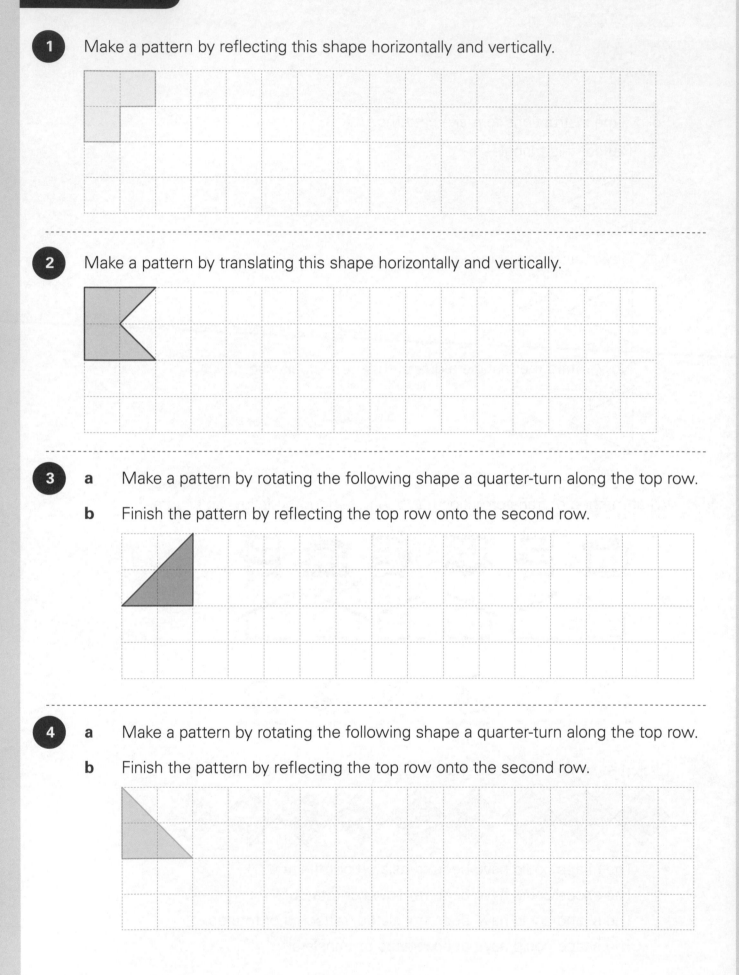

2   Make a pattern by translating this shape horizontally and vertically.

3   **a**   Make a pattern by rotating the following shape a quarter-turn along the top row.

    b   Finish the pattern by reflecting the top row onto the second row.

4   **a**   Make a pattern by rotating the following shape a quarter-turn along the top row.

    b   Finish the pattern by reflecting the top row onto the second row.

OXFORD UNIVERSITY PRESS

Symmetrical faces—spot the difference

1 Do you have a symmetrical head? A lot of people think they do, but not many people's heads are truly symmetrical.

Look at the two pictures of the woman's head. In the first photo, her head might appear symmetrical, but compare it with the second photo.

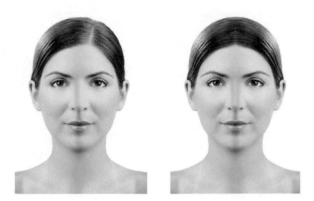

In the second photo, the left-hand side of the woman's face has been copied, reflected and pasted onto the right-hand side.

Identify the feature(s) in the first picture that make the woman's head asymmetrical.

_____

_____

2 Use the grid to help you to reflect the drawing of the left part of a head. Make the drawing as symmetrical as you can.

3 Find a picture you like of something that is symmetrical or near-symmetrical. Cut it in half and draw the reflected part of the image.

Using grid lines can help you with the accuracy of your drawings.

The Cartesian coordinate system

Practice

1 Give the coordinate points for the:

a blue dot

_____

b purple dot

_____

c green dot

_____

d yellow dot

_____

e red dot

_____

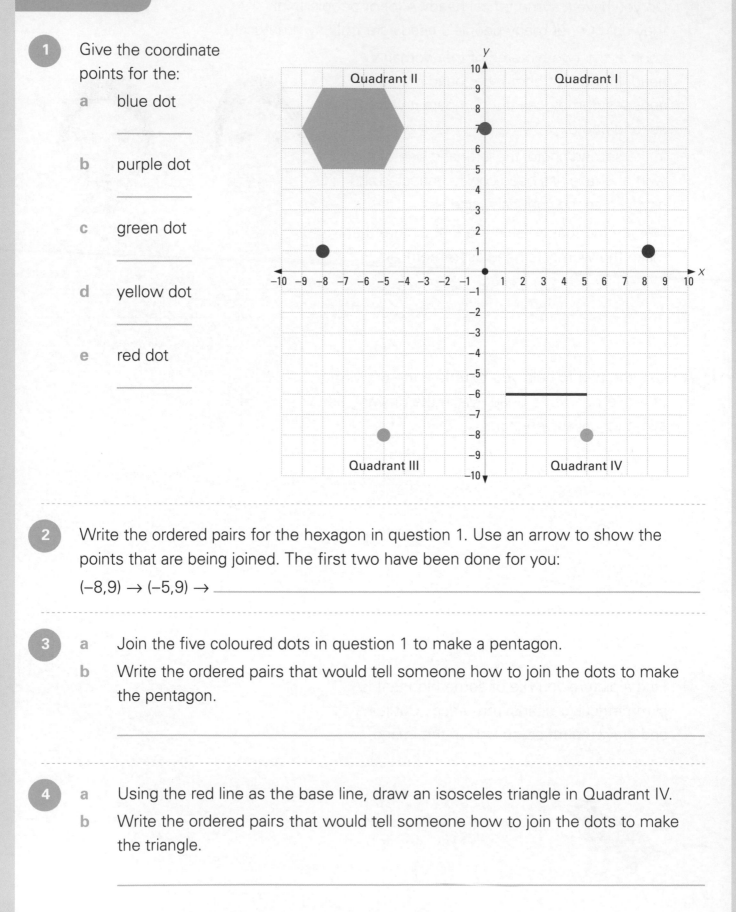

2 Write the ordered pairs for the hexagon in question 1. Use an arrow to show the points that are being joined. The first two have been done for you:

$(-8,9) \rightarrow (-5,9) \rightarrow$ _____

3 a Join the five coloured dots in question 1 to make a pentagon.

b Write the ordered pairs that would tell someone how to join the dots to make the pentagon.

_____

4 a Using the red line as the base line, draw an isosceles triangle in Quadrant IV.

b Write the ordered pairs that would tell someone how to join the dots to make the triangle.

_____

OXFORD UNIVERSITY PRESS

Challenge

1 **a** Write the ordered pairs that would tell someone how to join the dots to draw the aeroplane shape in Quadrant II.

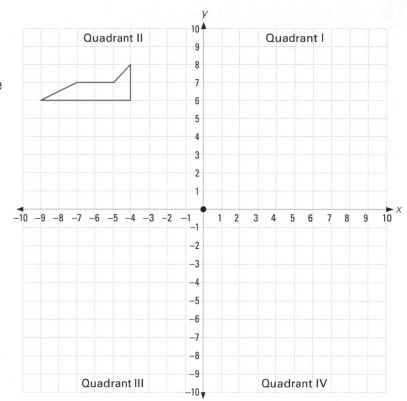

_____

b Reflect the plane into Quadrant I so that the front tip is at (9,6).

c Write the ordered pairs that would tell someone how to join the dots to draw the aeroplane in Quadrant I.

_____

2 Follow the ordered pairs to draw the left-hand side of a space rocket onto the grid above.

(0,10) → (–1,9) → (–1,3) → (–2,2) → (–2,–4) → (–4,–5) → (–4,–9) →
(–2,–8) → (–1,–7) → (0,–7)

_____

3 **a** Reflect the drawing from question 2 into Quadrants I and IV.

b Write the ordered pairs that would tell someone how to join the dots to draw the right-hand side of the rocket.

_____

1 On this number plane we can see that the points that the red line passes through are all in Quadrant I. Using ordered pairs, we could write:

$(0,10) \rightarrow (1,8) \rightarrow (2,6) \rightarrow (3,4)$ and so on.

A different way is to use x for the position on the x axis and y for the position on the y axis. We can then write the coordinate points in a table of values:

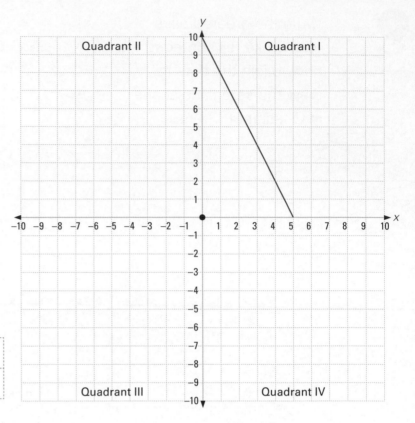

x	0	1	2	3	4	5
y	10	8	6			

Complete the table of values.

---

2 In order to write a rule for the table of values in question 1, we could say:

To find the position of y you start with 10 and subtract double the value of x.

This rule might sound very complicated, but we can turn it into a simple equation:

$y = 10 - (x \times 2)$.

You can test the rule by using 3 as the value of x: $y = 10 - 3 \times 2$.

So, remembering the order of operations, this means the position of y should be 4, and this is correct.

a Using this rule, continue the table of values as far as $x = 10$

x	0	1	2	3	4	5	6	7	8	9	10
y	10	8	6								

b Into which quadrant does the line pass? _____

c Continue the red line using the new table of values.

---

3 For a different line we need a different equation.

a Using $y = -10 + (x \times 2)$, complete the table of values:

x	0	1	2	3	4	5	6	7	8	9	10
y	-10	-8									

b Plot the points on the number plane above and draw a line to connect them.

OXFORD UNIVERSITY PRESS

UNIT 9: TOPIC 1
Collecting, representing and interpreting data

Practice

1 Class 6G looked after the school garden. They had an open week. They made a frequency table and then a pictograph to show the number of visitors that came.

   a Use the information to fill in the gaps in the frequency table and in the pictograph.

Frequency table: The number of people who came to visit the school garden					
Day:	Monday	Tuesday	Wednesday	Thursday	Friday
Number:		14	21		

   b They gave each visitor a flower. How many flowers did they give away?

_____

Graph to show the number of visitors to our garden in open week.

 Key = 4 people

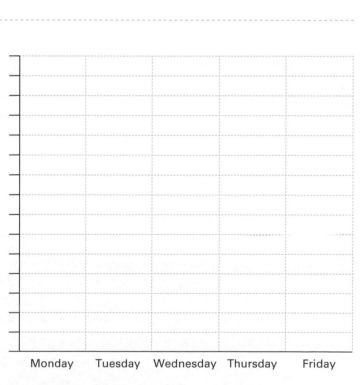

Monday   Tuesday   Wednesday   Thursday   Friday

---

2 **a** Transfer the data from question 1 onto a line graph. Decide on a suitable scale for the graph.

   b Write a title on the graph and label the vertical axis.

   c What trend about the number of visitors does the line graph show?

_____

_____

_____

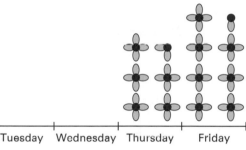

Monday   Tuesday   Wednesday   Thursday   Friday

   d If the garden were open for a sixth day, what would be your prediction for the number of visitors? Give a reason for your answer.

_____

_____

Challenge

1 The populations of five countries are compared on this circle graph.

Comparing the populations of 5 countries

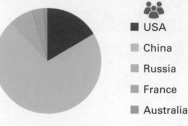

- ■ USA
- ■ China
- ■ Russia
- ■ France
- ■ Australia

   a Which country has a population that is almost three quarters of the total population of the five countries? _____

   b Write a statement of finding from the information in the graph.

_____

2 The same data is shown on this bar graph.

   a One advantage of a circle graph is that it is an easy way of comparing data. What is an advantage of presenting the same data on a bar graph?

_____

_____

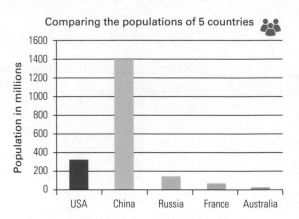

   b From the graph we can see that the population of China is about 1400 million. Write the population in figures. _____

   c Estimate the population of the USA. _____

3 This graph compares the populations of pet cats in the same countries. Write a statement of finding about the pet cat population of the USA.

_____

_____

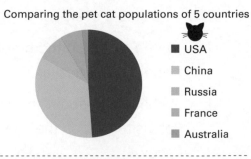

Comparing the pet cat populations of 5 countries

- ■ USA
- ■ China
- ■ Russia
- ■ France
- ■ Australia

4 This bar graph also represents the data for the pet cat populations.

   a From the data we can see that the USA appears to be a nation of cat lovers. If someone made a statement of finding that, "Around one person in four in the USA has a pet cat", how accurate would that statement be?

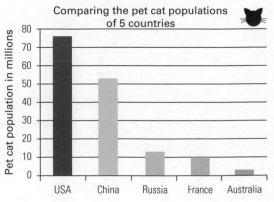

OXFORD UNIVERSITY PRESS

1 Making a graph is a relatively simple method of representing data in an interesting way. For example, the table below shows the countries that produce the most gold. Either with pen and paper or by using a program such as Excel, this information can quickly be turned into a graph.

The top gold-producing countries	
Country	**kg of gold per year**
China	362 000 kg
Australia	258 000 kg
United States	234 000 kg
Russia	199 442 kg
South Africa	181 000 kg

The world's top 5 gold-producing countries

a Sometimes it is necessary to round figures to make the data easier to handle. Which gold-production figure do you think would need to be rounded and how would it be rounded?

b Notice that the figures on the vertical axis do not show the number in full—they say 400, 350 instead of 400 000, 350 000 etc. Why do you think this is?

2 Use some of the data in the following tables to make a graph.

Countries with the most threatened mammals		
Country	**In 2010**	**In 2015**
Indonesia	183	186
Madagascar	63	118
Mexico	99	101
India	93	96
Brazil	79	81

Top banana-producing countries	
Country	**Bananas (tonnes)**
India	27 575 000 t
China	12 075 238 t
Philippines	8 645 749 t
Brazil	6 892 622 t
Ecuador	5 995 527 t

Cities with the most skyscraper buildings		
City	**Skyscrapers**	**Population**
Hong Kong, China	1268	6 980 412
New York City, USA	600	8 274 527
Tokyo, Japan	411	9 262 046
Chicago, USA	295	4 553 009
Dubai, UAE	249	10 356 000

Countries with the most tourists per year	
Country	**Number of tourists**
France	83.01 million
USA	66.96 million
China	57.72 million
Spain	56.70 million
Italy	46.36 million

Practice

1 Students in a Year 6 class wanted to be allowed to use smartphones in class. They showed their teacher the results of a survey and said that it was clear that most people thought it was a good idea.

Should students use smartphones in class?

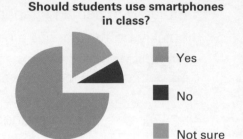

Yes

No

Not sure

  a About what percentage of people thought that smartphones should be allowed in class? _____

  b 24 people were surveyed. Four replied, "Not sure". Estimate the number who answered, "No". _____

2 The teacher asked who had been surveyed. The students answered that all the students in the class had been surveyed. Was the graph based on primary or secondary data?

_____

3 The teacher said that they should survey the students' parents, but suggested that parents would probably think the smartphones would be used for playing games. Write a survey question that would be suitable for surveying the parents.

_____

4 The teacher asked a few students to do some research to find out what other people thought about the idea.

  a Circle the most suitable subject to search for online:

   • *Should smartphones be used in classrooms?*

   • *Should smartphones be banned in classrooms?*

   • *Should Year 6 students be allowed to use smartphones in classrooms?*

  b Give a reason for your choice.

_____

5 The result of the survey of the parents was made into a graph. Estimate the percentages for *No* and *Yes*.

Do parents think students should use smartphones in class?

_____

Yes

No

OXFORD UNIVERSITY PRESS

1 The teacher of the Year 6 class on page 95 asked what survey question the students had used. It was, "Should we be able to use the calculator function on our smartphones in class?" Comment on the appropriateness of the survey question.

_____

2 A local newspaper heard about the graph and published it with the headline:

SHOCKING NEWS:
Most parents want their children to use smartphones in class!

One of the Year 6 parents angrily contacted the newspaper and said that the story was completely untrue. Was it? Circle one response below. The headline was:

- based on fact and completely true
- based on fact and partly true
- completely untrue
- based on fact and possibly true.

3 The newspaper editor decided to do some research of her own. She asked people to respond by calling a survey line.

> Should students be allowed to play with their smartphones in class? Press 1 for *Yes* and 2 for *No*. Call NOW.

Was the editor's survey a census survey or a sample survey? _____

4 A reporter did some research and read that, after smartphones were banned at some schools in England, test scores for students aged 16 rose by 6.4 per cent. He published an article with the headline:

> Recent survey shows that students do better when smartphones are banned!

Comment on the fairness of the article.

_____

_____

5 The result of the newspaper editor's telephone poll was published the next day.

a 102 people answered the survey. Estimate the numbers who responded *Yes* and *No*. _____

b Comment on the difference between the newspaper editor's survey question (question 3) and the graph title. _____

_____

_____

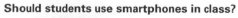

Should students use smartphones in class?

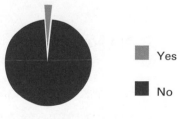

Yes

No

1 There are different opinions about when and if children should be allowed to own a smartphone.

A British newspaper reported in 2017 that billionaire Microsoft founder, Bill Gates, "… banned his children from mobile phones until they turned 14".

Others believe that smartphones are a good idea for children, because they keep students happy.

a What might be another reason why students should be allowed to use a smartphone?

_____

b Why do you think Bill Gates believes that children under 14 should not have a smartphone?

_____

2 In June 2017, SBS TV reported that "… one third of pre-schoolers own smartphones". What is your opinion about pre-schoolers having their own smartphones?

_____

_____

3 Whatever a person's viewpoint, the fact is that there has been a rise in smartphone ownership by children since 2010.

a Estimate the percentage of children who owned a smartphone in 2010.

_____

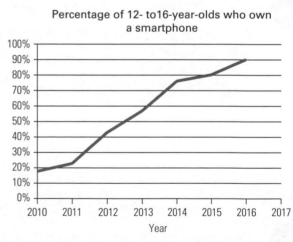

Percentage of 12- to16-year-olds who own a smartphone

b The line graph shows a definite trend in smartphone ownership by children. Comment on the trend and give your opinion of the reason for it.

_____

_____

_____

c No figures were available for 2017. What is your estimate for the percentage?

_____

d With your teacher's permission, carry out some research to find out about smartphone ownership in your class. Decide how best to represent your data.

Practice

1   Look at the data set and draw lines to match the numbers with the correct measures of averages.

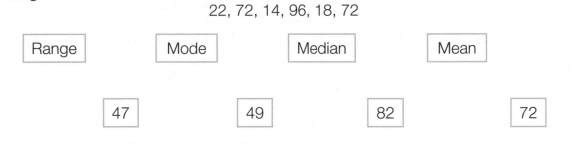

22, 72, 14, 96, 18, 72

| Range | Mode | Median | Mean |

| 47 | 49 | 82 | 72 |

2   The table below shows how many boxes of chocolates were sold for the school fundraiser.

Class	No. of boxes
Foundation	71
Year 1	35
Year 2	31
Year 3	30
Year 4	35
Year 5	68
Year 6	73

a   Estimate whether you think the mode, the mean or the median will be greater.

_____

b   Find the mode. _____

c   Calculate the mean. _____

d   Find the median. _____

e   Write in your own words what a range is and how you would find it for this data set.

_____

_____

_____

When might we need to know the mean, median and mode in real life?

Challenge

1 Find the missing number in each data set if:

a the mean is 7.

8, 6, 9, 5, _____

b the mean is 20.

25, 16, 14, _____

c the mode is 100.

74, 100, 82, 82, 67, 100, _____

d the mode is 42.

42, 3, 4, 50, 42, 43, _____

e the median is 10.

12, 7, 9, 5, 16, 12, _____

f the median is $5\frac{1}{2}$.

1, 5, 2, 10, 8, 9, 2, _____

g the range is 12.

10, 15, 20, 17, 14, _____

h the range is 100.

28, 117, 93, 32, 46, 21, _____

---

2 A Year 6 class had a paper plane throwing contest. The table below shows their results. You may use a calculator where required to help you find:

Name	Distance
Hannah	24 m
Jordan	17 m
Jimin	31 m
Chiara	26 m
Omar	25 m
Lewis	19 m
Natalie	33 m
Salma	26 m

a the mean distance. _____

b the median distance. _____

c the mean of the four longest distances. _____

d the mean of the four shortest distances. _____

e the range of the four longest distances. _____

f the mode. _____

---

3 Jason's spelling scores for the term were:

24, 21, 13, 20, 21, 25, 23, 21, 24, 22

a What is his mean score? _____

b What is the median? _____

c What is the mode? _____

d What is the range? _____

e If Jason scored 13 on each of his next two tests, what would his mean score be? _____

OXFORD UNIVERSITY PRESS

1 Isabelle scored points in the first 10 basketball games of the season.

a If the mode was 12, what might her score in each game have been? Show two sets of options.

b If her median score was 10, what might her score in each game have been? Show two sets of options.

c If her mean score was 13, what might her score in each game have been? Show two options.

d Write the range for your answers to questions 1 to 3.

Practice

1

a The spinner has a 1-in-how-many chance of landing on white? _____

b What chance does the spinner have of landing on green? Answer using a percentage. _____

c What chance does the spinner have of **not** landing on green? Answer using a decimal. _____

d There is half as much chance of the spinner landing on yellow as on green. Describe the percentage chance for yellow. _____

e For which colour is there a 0.375 chance? _____

2 Describe something in your school that has the following chance of happening:

a 2% chance _____

b 50–50 chance _____

c 0% chance _____

d 0.99 of a chance _____

e $\frac{1}{4}$ of a chance _____

3 Which of these describes the chance of this spinner landing on red? Circle any that are appropriate.

$\frac{1}{4}$ $\frac{1}{5}$ 20% 25% 0.25 0.2 0.5 1 in 5 $\frac{2}{10}$

4 A coach says the team has a very good chance of winning a game. Give a percentage value for the chance.

_____

Challenge

1 **a** Colour this spinner so that the following probabilities are true:

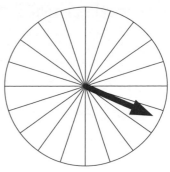

- There is a 20% chance for green.
- There is 0.3 of a chance for red.
- There is a 5% chance for purple.
- There is $\frac{1}{5}$ of a chance for yellow.
- There is 0.15 of a chance for blue.

b Describe the probability of the spinner landing on the remaining white part of the spinner as a fraction, as a decimal and as a percentage. _____

---

2 Each jar below contains 60 jelly beans, of all colours. The number of black jelly beans in each jar is shown on the label. Write a value to show the probability of picking out a black jelly bean first time if you chose one without looking. Choose from the list.

$\frac{1}{5}$ 0.25 $\frac{2}{3}$ 75% 0.3̇ $\frac{7}{12}$

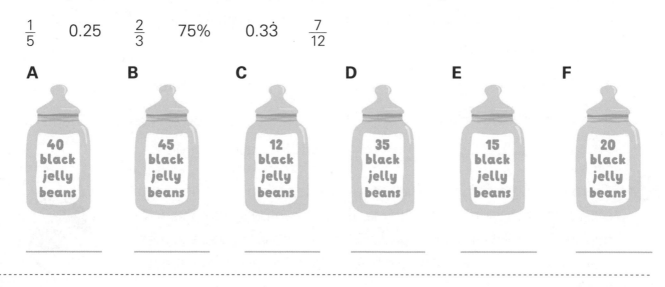

A	B	C	D	E	F
40 black jelly beans	45 black jelly beans	12 black jelly beans	35 black jelly beans	15 black jelly beans	20 black jelly beans

_____ _____ _____ _____ _____ _____

---

3 There is a mix of red and green beads in each bag of 50. Evie takes 20 from each bag without looking and records the results. Predict the number of red and green beads in each bag of 50.

Bag	After 20 have been taken out:		My prediction after 50 have been taken out:	
A	Red: 16	Green: 4	Red:	Green:
B	Red: 2	Green: 18	Red:	Green:
C	Red: 6	Green: 14	Red:	Green:
D	Red: 12	Green: 8	Red:	Green:

Bag A Bag B Bag C Bag D

In a normal pack of playing cards there are 4 types, called *suits*. There are 13 cards in each suit, making a total of 52. Each suit has the same cards.

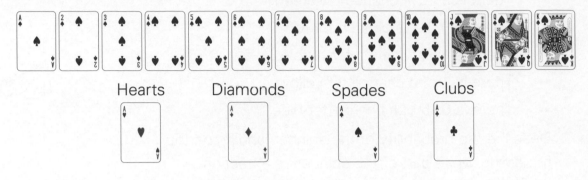

Hearts   Diamonds   Spades   Clubs

1 For the following tasks, decide whether the values are more appropriately expressed as fractions, decimals or percentages. Imagine that the cards are turned over and shuffled around, and that somebody has to choose one without looking. What is the chance of the person guessing correctly if they predict that the card will be:

a   a red card? _____   b   a number 2? _____

c   an odd numbered card (counting   d   a red 4? _____
    the ace as one)? _____

e   a diamond? _____   f   a picture card? _____

g   a picture card with a jack or a king?   h   a card lower than 5 (counting the
    _____   ace as one)? _____

2 There is a game called *Twenty-one* in which people try to get a total of 21 with their cards. In this game, the ace can be worth one or eleven. The quickest way to get to 21 is to choose an ace and then a ten. The picture cards are worth 10 each, so an ace plus a ten or a picture card makes 21.

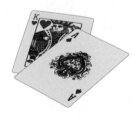

a   What is the chance of choosing an ace on the first try? _____

b   If you choose an ace on the first try, what is the chance of choosing a card worth 10 with your next pick? Remember that there won't be 52 cards left to choose from! _____

c   Express the chance of the second card being worth 10 as a percentage (to the nearest whole number). _____

3 Carry out some investigations by looking at the 52 cards. You could write some questions to give to others in the class, but make sure you know the answer yourself! For example, if you picked up two picture cards, what would be the probability of getting an ace to make 21 on the next try?

OXFORD UNIVERSITY PRESS

UNIT 10: TOPIC 2
Conducting chance experiments and analysing outcomes

You will need a pack of playing cards.

Card game 1: Finding the right queen

1 For this game you need the four queens from a pack of cards. Turn them face down and shuffle them around. The chance of picking out any queen is obviously 100%.

  a What is the percentage chance of choosing the Queen of Diamonds? _____

  b What is the decimal chance of choosing a red queen? _____

  c Write the chance of **not** choosing the Queen of Spades as a fraction. _____

2 **a** Still using only the four queens, choose a card at random 20 times. How many times would you expect to choose the queen from each suit? _____

  b What fraction of the number of times would that be for each suit? _____

  c Carry out the experiment. Record your results with tally marks in the table below.

Queen	Tally	Tally total	Write as a fraction of 20
Spades			
Clubs			
Hearts			
Diamonds			

  d How did your prediction compare with the actual results?

  _____

Card game 2: Kings and queens

3 For this game you will need four queens and four kings.

  a Repeat the experiment from question 2, choosing a card at random 20 times. Predict the number of times you will choose a king or a queen. _____

  b Carry out the experiment. Record the results with tally marks in the table below.

Card	Tally	Tally total	Write as a fraction of 20
A queen			
A king			

  c Was your prediction closer to the actual number than it was in question 2? Suggest a reason. _____

Challenge

Card game 3: Jack of Diamonds

For this game, you will need four queens and the Jack of Diamonds.

1 Write a probability value for choosing:

a queen _____ the Jack of Diamonds _____ a black queen _____

---

2 **a** Repeat the experiment from question 2 on the previous page, this time using the four queens and the Jack of Diamonds. How many times would you expect each of the following to appear?

a queen _____ a black queen _____ the Jack of Diamonds _____

b Repeat the experiment choosing from the five cards 20 times.

Card	Tally	Tally total	Write as a fraction of 20
A queen			
A black queen			
Jack of Diamonds			

c How did your results compare to those of a classmate?

_____

---

3 You need the following cards (these are pictured at right):

Hearts: 2, 3, 4 and 5   Spades: 6, 7 and 8

Clubs: 9 and 10   Diamonds: Jack of Diamonds

Write the percentage chance for choosing:

a a heart _____ **b** a spade _____

c a club _____ **d** a diamond _____

e a red card _____ **f** an even numbered spade _____

---

4 If you were to add the Ace of Diamonds to the ten cards in question 3, would the chance of choosing the following be higher or lower than it is in question 3?

a a heart _____ **b** a spade _____

c a club _____ **d** a diamond _____

e a red card _____ **f** the Jack of Diamonds _____

---

5 Write the chance for each outcome in question 4 as a fraction, a decimal (to 2 decimal places) and a percentage. _____

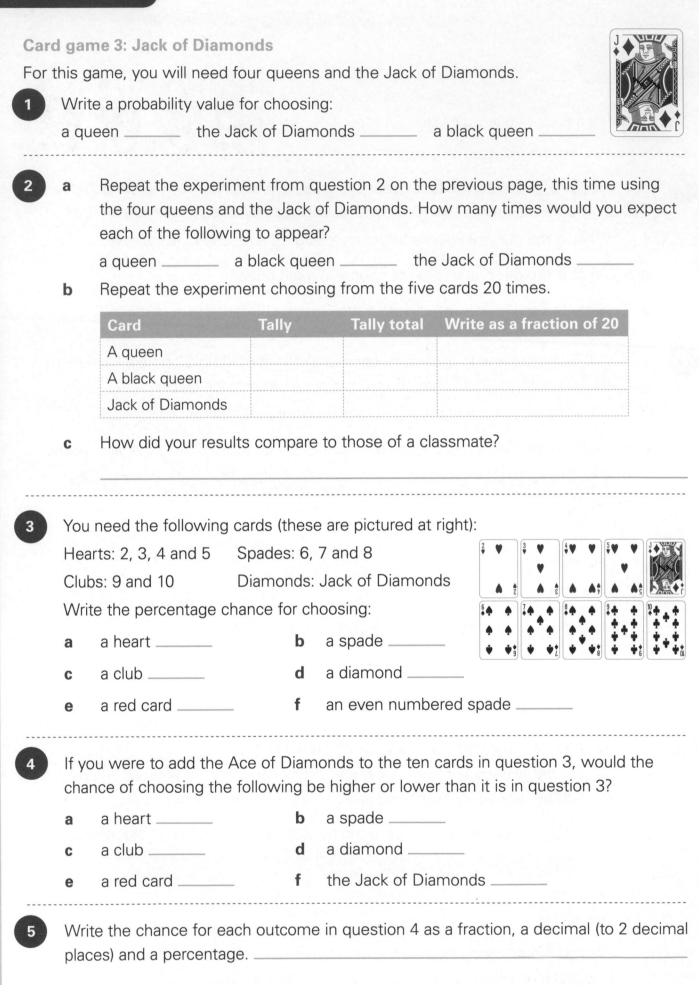

OXFORD UNIVERSITY PRESS

Card game 4: Jumps and slides

1 **You need:**

Two or more players, a full pack of 52 cards, a pencil and paper for each player.

The game:

The aim of the game is to get to 100.

You could use a hundred grid for each person to keep track.

Rules:

The cards are shuffled and placed upside-down in the middle. Players take turns to pick up a card and jump forward by the number on the card.

An ace counts as 1 or 11. Queens and kings count as 15.

However, there are **jumps** and **slides**:

- Any card that is a multiple of 5: jump forward an extra 5 steps.
- Any card that is a multiple of 3: slide back 3 steps.
- If you land on a prime number: slide back to the previous prime number.
- Any ace: jump forward an extra 10 steps.
- Jack of Hearts, Clubs or Spades: slide back 10 steps (or to zero).
- Jack of Diamonds is the **Star Card**: jump straight to 100—you win!

If you decide to change the rules, make sure all players agree. For example, if you are up to 96, do you have to get exactly 4 to get to 100, or will any number higher than 3 allow you to win?

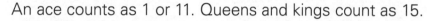

2 Make up your own game using playing cards. In the game above, there was only a 1-in-52 chance of getting the Star Card. How could you increase the chances of getting a Star Card? When you have invented your game, write out the rules and then try it out to see if the rules work.

ANSWERS

UNIT 1: Topic 1 Place value

Practice

1 a 700 000   b 50 000
   c 6 000 000   d 10 000 000
   e 800   f 200 000 000

2 a seven hundred and thirty-four thousand, eight hundred and fifteen
   b sixty-two million, seven hundred and fifty-nine thousand, three hundred and seventy-seven
   c six million, two hundred and nineteen thousand, seven hundred and thirty
   d thirteen million, five hundred and eighty-five thousand, one hundred and four
   e five million, four hundred and eighty-seven thousand, eight hundred
   f two hundred and eighteen million, eight hundred and nineteen thousand, nine hundred and ninety-nine

3 a 413 629 = 400 000 + 10 000 + 3000 + 600 + 20 + 9
   b 3 746 123 = 3 000 000 + 700 000 + 40 000 + 6000 + 100 + 20 + 3
   c 52 065 350 = 50 000 000 + 2 000 000 + 60 000 + 5000 + 300 + 50
   d 43 200 806 = 40 000 000 + 3 000 000 + 200 000 + 800 + 6
   e 680 405 020 = 600 000 000 + 80 000 000 + 400 000 + 5000 + 20

Challenge

1 a 10 235 789
   b 19 875 320
   c 123 578.9 (The zero is 'non-significant'.)
   d 53 210.987

2 a 31 084 509 matches to spike abacus B.
   b Three hundred and ten million, eighty thousand, four hundred and fifty-nine matches to spike abacus D.
   c 310 084 509 matches to spike abacus A.
   d Three hundred and one million, eight hundred and four thousand, five hundred and nine matches to spike abacus C.

3 a $395 000
   b $1 300 000
   c $910 000
   d $1 500 000

Mastery

1 Students should circle the following:
   3 878 224   3 949 999
   3 898 100   3 918 885

2 A range of correct answers is possible. Students will demonstrate a comprehensive understanding of place value, by giving a number of options, including:
   • 22 000 000 + 900 000 + 8000 + 500 + 40 + 5
   • 20 000 000 + 2 000 000 + 900 000 + 8000 + 545
   • 20 000 000 + 2 000 000 + 900 000 + 8000 + 500 + 40 + 5
   • 22 000 000 + 900 000 + 8000 + 500 + 45
   • 20 000 000 + 2 000 000 + 900 000 + 8000 + 500 + 45

3 There are 24 possible answers. Look for students who recognise that the 2 digit must be in the millions place, the 3 digit in the hundreds-of-thousands place and that, in order to round to the nearest hundred-thousand, the tens-of-thousands digit must be lower than 5.

Look also for those who adopt a systematic approach to choosing numbers, e.g.

2 315 679	2 316 579	2 317 569	2 319 567
2 315 697	2 316 597	2 317 596	2 319 576
2 315 769	2 316 759	2 317 659	2 319 657
2 315 796	2 316 795	2 317 695	2 319 675
2 315 967	2 316 957	2 317 956	2 319 756
2 315 976	2 316 975	2 317 965	2 319 765

UNIT 1: Topic 2 Square numbers and triangular numbers

Practice

1 Student shades:
   1, 4, 9, 16, 25, 36, 49, 64, 81 and 100.

2 There is an equal number of odd and even square numbers.

3 Students circle 5×5 and 5^2.

4 Students circle 1, 3, 6, 10, 15, 21, 28, 36, 45, 55.

5 1 and 36.

Challenge

1 $1 + 3 + 5 + 7 + 9 + 11 + 13 + 15 + 17 + 19 + 21$

2 a 169
   b $1 + 3 + 5 + 7 + 9 + 11 + 13 + 15 + 17 + 19 + 21 + 23 + 25$

3 11

4

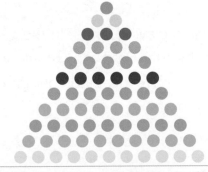

$1 + 2 + 3 + 4 + 5 + 6 + 7 + 8 + 9 + 10 + 11 = 66$

Mastery

1 a 3rd TN = $3 \times (3 + 1) \div 2 = 6$
   b 5th TN = $5 \times (5 + 1) \div 2 = 30 \div 2 = 15$

2 Answers will vary depending on the level of motivation of the students and/or the teacher.

   20th TN = $20 \times (20 + 1) \div 2 = 420 \div 2 = 210$

   Students could be encouraged to use a calculator for larger triangular numbers, such as the 60th (1830) and 100th (5050) triangular numbers.

UNIT 1: Topic 3 Prime and composite numbers

Practice

1

Number	Factors (numbers it can be divided by)	How many factors?	Is it prime or composite?
21	1, 21, 3, 7	4	composite
22	1, 22, 2, 11	4	composite
23	1, 23	2	prime
24	1, 24, 2, 12, 3, 8, 4, 6	8	composite
25	1, 25, 5	3	composite
26	1, 26, 2, 13,	4	composite
27	1, 27, 3, 9	4	composite
28	1, 28, 2, 14, 4, 7	6	composite
29	1, 29	2	prime
30	1, 3, 10, 30, 2, 15, 5, 6	8	composite

2 The factors of 12 are 1, 12, 2, 6, 3 and 4.

3 a 28   b 24 and 30   4 2

5 a 25 is the square number on the chart. It has 3 factors.
   b 4 or 9

Challenge

1 a 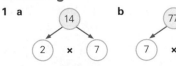   b

The prime factors of 14 are 2 and 7. So, $2 \times 7 = 14$

The prime factors of 77 are 7 and 11. So, $7 \times 11 = 77$

c

The prime factors of 30 are 2, 3 and 5. So, $2 \times 3 \times 5 = 30$

OXFORD UNIVERSITY PRESS

2 a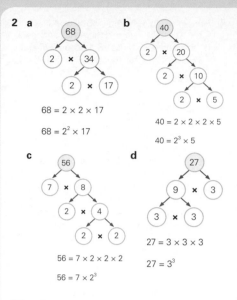

68 = 2 × 2 × 17

68 = 2^2 × 17

b

40 = 2 × 2 × 2 × 5

40 = 2^3 × 5

c

56 = 7 × 2 × 2 × 2

56 = 7 × 2^3

d

27 = 3 × 3 × 3

27 = 3^3

Mastery

1 Answers may vary, but students will demonstrate mastery by explaining that 37 has only two factors, 1 and itself.

2

	Prime factors	Index form	Number
a	2 × 2 × 2 × 3	2^3 × 3	24
b	2 × 2 × 2 × 5 × 5	2^3 × 5^2	200
c	2 × 2 × 2 × 2 × 5	2^4 × 5	80
d	2 × 2 × 3 × 7	2^2 × 3 × 7	84

3 a 4 **b** 2 **c** 7 **d** 3
4 a 6 **b** 8 **c** 9
5

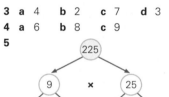

225 = 3 × 3 × 5 × 5

225 = 3^2 × 5^2

$\sqrt{225}$ = 3 × 5 = 15

Check: 15 × 15 = 225

UNIT 1: Topic 4 Mental strategies for addition and subtraction

Practice

1 a 658 **b** 659 **c** 1358
   d 2459 **e** 4999 **f** 12 989
2 a 239 **b** 729 **c** 1060
   d 267 **e** 567 **f** 1401
3 Teachers may prefer to ask students to compare the strategies they used with a partner or other group members. Examples of strategies that might be used include:
   a "I added 350 to 551 and then took away 2." Answer: 899
   b "I took 300 away and then added 1 back." Answer: 1173

c "I added 350 and 350 and then added the 700 to 8000." Answer: 8700
d "I took away 500 from 24 746 = 24 246. Then I took away another 20 = 24 226." Answer: 24 226

Challenge

1 Students could be asked to explain to each other the way that they chose to round the amounts and to justify their chosen method of rounding. Examples of the way the amounts might be rounded include:

	Price	Rounded price	I rounded to the nearest ...
a	$450 300 000	$450 000 000	million
b	$179 000 000	$180 000 000	ten million
c	$170 400 000	$170 000 000	million
d	$142 400 000	$140 000 000	ten million
e	$119 922 500	$120 000 000	million

2 a $593M (Actual amount = $592 700 000) **b** $8 600 000
   c $77 500 **d** $312 800 000

3 Here are some examples of ways that rounding could be used to estimate the answers, and the exact answers:

	Problem	Round the numbers	Estimate the answer	Calculator answer
a	3085 + 1877	3000 + 2000	5000	4962
b	6018 – 3119	6000 – 3000	3000	2899
c	35 056 + 4023	35 000 + 4000	39 000	39 079
d	29 856 – 20 028	30 000 – 20 000	10 000	9828
e	400 247 + 49 588	400 000 + 50 000	450 000	449 835
f	398 108 – 197 434	400 000 – 200 000	200 000	200 674
g	5 498 858 – 2 889 917	5 500 000 – 3 000 000	2 500 000	2 608 941

Mastery

1 a $2600
   b $2605
2 a Look for students who demonstrate an understanding of the connection between addition and subtraction. Teachers will probably wish to encourage students to use a calculator to subtract 123456789 from 987654321, and they will arrive at an answer of 864197532.
   b Students who have a good understanding of place value will write the answer as 864 197 532. In words this is eight hundred and sixty-four million, one hundred and ninety-seven thousand, five hundred and thirty-two.
3 Look for students who demonstrate time-saving strategies, such as looking for addition pairs that total 1000. Possible solutions are:

Performance	Number of people
1	499
2	501
3	498
4	502
5	497
6	503
7	496
8	504
Total	4000

UNIT 1: Topic 5 Written strategies for addition

Practice

1 a 223 **b** 456
   c 789 **d** 1234
   e 12 345 **f** 123 456
   g 1 234 567 **h** 12 345 678
   i 123 456 789

Challenge

1 a

	Country	Length of paved roads (km)	Length of unpaved roads (km)	Total length of roads (km)
1	USA	4 165 110	2 265 256	6 430 366
2	India	1 603 705	1 779 639	3 383 344
3	China	1 515 797	354 864	1 870 661
4	France	951 220	0	951 220
5	Japan	925 000	258 000	1 183 000
6	Russia	738 000	133 000	871 000
7	Spain	659 629	6663	666 292
8	Canada	415 600	626 700	1 042 300
9	UK	388 008	0	388 008
10	Australia	336 962	473 679	810 641
11	Brazil	96 353	1 655 515	1 751 868

b USA, India, China, Brazil, Japan, Canada, France, Russia, Australia, Spain, UK

ANSWERS

Mastery

In this section look for students who use their knowledge of place value and who see the connection between addition and subtraction as part of their problem-solving strategies.

1. Students could subtract any 5-digit number from 123 321, finding two addends that would give the correct total. For example, 123 321 – 52 897 = 70 424, so the two addends would be 52 897 + 70 424 = 123 321. A similar process could be used for the other two sums. Those looking for a short cut to find the other two solutions could subtract one from one addend and add one to the other to solve the problem (i.e. 52 896 + 70 425 and 52 895 + 70 426).

2. This problem can be solved in many ways. Perhaps the easiest would be to use a knowledge of expanded numbers, renaming 123 456 789 as:

   123 000 000 + 400 000 + 50 000 + 6789.

3. The question does not specify the number of addends, but students will probably not choose to have a high number. They will probably use a process of trial and error to solve the problem. Look for students who see that, since the ones column should total 6, one of the addends will end in 5 and another in 1. One possible solution is 10 755 + 1171 + 120.

4. As with question 1, students who have a good grasp of place value will probably use their knowledge of expanded numbers to solve the problem. For example, they could start with the answer and subtract 21 000 000 from it, leaving 966 654 321. 21 000 000 could then be expanded into 19 000 000 + 1 900 000 + 100 000. The resulting addition would be 966 654 321 + 19 000 000 + 1 900 000 + 100 000 = 987 654 321.

UNIT 1: Topic 6
Written strategies for subtraction

Practice

1.
a	121 212	b	232 323
c	343 434	d	454 545
e	565 656	f	676 767
g	787 878	h	898 989
i	989 898	j	878 787

2. Students who see the connection between subtraction and addition will solve the problem by adding 767 676 and 343 965 to find that the number on the top line is 1 111 641.

Challenge

a	321 321	b	432 432
c	43 214 321	d	54 325 432
e	65 436 543	f	76 547 654
g	87 658 765	h	98 769 876
i	123 454 321	j	543 212 345

2. There are many different solutions that are possible. Students will probably use trial and error, beginning by writing a top line with six zeros, e.g. 300 400 700. They can then subtract 123 456 789 from the number to work out the number that needs to appear on the second line of the subtraction.

In this example the subtraction would become 300 400 700 – 176 943 911 = 123 456 789.

Mastery

1. 6906 tonnes
2.
   a. Mexico, Chile and Colombia
   b. 35 322 t
   c. 20 139 t
   d. 953 175 t
   e. 1 142 892 t – 829 099 t = 313 793 t
   f. 1 971 991 t – 1 758 549 t = 213 442 t

UNIT 1: Topic 7
Mental strategies for multiplication and division

Practice

1.
		×10	×100	×1000
a	43	430	4300	43 000
b	87	870	8700	87 000
c	235	2350	23 500	235 000
d	$7.75	$77.50	$775.00	$7750.00
e	250	2500	25 000	250 000
f	36	360	3600	36 000
g	$0.25	$2.50	$25.00	$250.00
h	$10.15	$101.50	$1015.00	$10 150.00

2.
		÷10	Write the multiplication-fact partner
a	350	35	35 × 10 = 350
b	3600	360	360 × 10 = 3600
c	5000	500	500 × 10 = 5000
d	$12.50	$1.25	$1.25 × 10 = $12.50
e	73	7.3	7.3 × 10 = 73

3.
		÷100	Write the multiplication-fact partner
a	800	8	8 × 100 = 800
b	$255	$2.55	$2.55 × 100 = $255
c	7000	70	70 × 100 = 7000
d	15 000	150	150 × 100 = 15 000
e	7250	72.5(0)	72.5 × 100 = 7250

Challenge

1.
		× 10	× 20	× 40	× 80
a	11	110	220	440	880
b	14	140	280	560	1120
c	21	210	420	840	1680
d	35	350	700	1400	2800
e	60	600	1200	2400	4800

2.
		÷ 10	÷ 20	÷ 40	÷ 80
a	800	80	40	20	10
b	4000	400	200	100	50
c	720	72	36	18	9
d	12 000	1200	600	300	150
e	4800	480	240	120	60

3.
a	500	b	960	c	750
d	560	e	750	f	2500
g	1500	h	1680	i	380
j	$65	k	$55	l	112
m	90	n	360		

4. Answers may vary, e.g. "I multiplied 22 × 10 = 220. Then I doubled 22 and doubled it again to get 88. I added 220 and 88 = 308. So, 22 × 14 = 308."

Mastery

1. 90 × 70 = 6300
2.
   a. Answers may vary. Look for students who see that, for numbers that are multiples of 10, the number of zeros in the answers is the same as the total number of zeros in the two numbers being multiplied.
   b. The investigation should result in proving the above to be correct. Some students may reach this conclusion after a couple of tries whereas others may wish to try many multiplications to test the theory.
3.
   a. The sentence that would not give the correct answer is, "You could halve 648 and move the digits one place bigger".
   b. Teachers may wish to give students the opportunity to share their choice of strategy with others so that students can see that what works best for one is not necessarily the choice of others.
4. This could form part of a cooperative group activity.

   5 for $107.50 = $21.50 per seat.
   32 for $624 = $19.50 per seat.

   Students will probably conclude that, even though only 30 seats are required, the school would save $21 by purchasing 32 seats from the second supplier.

UNIT 1: Topic 8
Written strategies for multiplication

Practice

1.
a	852	b	957	c	680
d	738	e	6956	f	7710
g	9380	h	11 459	i	13 968
j	19 215	k	32 040	l	56 781
m	14 560				

2. $18 468
3. 6 × $349 845 = $2 099 070 plus $337 845 = $2 436 915 in total

Challenge

1.
a	520	b	680	c	810
d	2120	e	1620	f	2200
g	5190	h	15 180	i	48 580
j	81 400	k	155 700	l	232 470

2.
a	600	b	910	c	864
d	3792	e	20 020	f	14 256

Mastery

1. Look for students who choose to make the calculations simpler by converting the masses to tonnes (e.g. the northern right whale has a mass of 77.7 t).
   a. The bowhead whale is almost twice the mass of the sperm whale. Bowhead whale = 86 000 kg; sperm whale = 43 700 kg (43 700 kg + 43 700 kg = 87 400kg).

OXFORD UNIVERSITY PRESS

b 1 787 100 kg or 1787.1 t

c 502.5 m

d 26 fin whales = 1 648 400 kg (1648.4 t); 12 blue whales = 1 644 000 kg (1644.0 t). Therefore, 26 fin whales have a greater mass than 12 blue whales.

e 15 × 77.7 t = 1165.5 t (1 165 500 kg) 15 × 43.7 t = 655.5 t (655 500 kg) The difference is 510 t (510 000 kg).

2 Teachers might encourage students to write multiplications involving more than two digits or ask them to use a calculator. By writing a long multiplication for 111 × 111, for example, we can observe the way the pattern is made:

				1	1	1
		×		1	1	1
				1	1	1
			1	1	1	0
+	1	1	1	0	0	
	1	2	3	2	1	

The pattern continues for 1111 × 1111 (=1 234 321), but calculators may not be available that will handle very large numbers. The pattern eventually breaks down after 111 111 111². However, students will probably have had enough by that stage!

Other patterns can be found by multiplying by 11, 111 and by 1111 instead of squaring the numbers. For example, 11 × 11 = 121, 111 × 11 = 1221, 1111 × 11 = 12 221.

UNIT 1: Topic 9 Written strategies for division

Practice

1 a 3 r1 or $3\frac{1}{4}$ **b** 7 r3 or $7\frac{3}{4}$

c 4 r1 or $4\frac{1}{8}$ **d** 7 r1 or $7\frac{1}{9}$

e 9 r3 or $9\frac{3}{8}$ **f** 8 r3 or $8\frac{3}{7}$

g 8 r2 or $8\frac{2}{9}$ **h** 9 r5 or $9\frac{5}{6}$

2 a 141 r3 and $141\frac{3}{4}$

b 74 r2 and $74\frac{2}{5}$

c 34 r2 and $34\frac{2}{6}$ ($34\frac{1}{3}$)

d 106 r3 and $106\frac{3}{7}$

e 321 r2 and $321\frac{2}{8}$ ($321\frac{1}{4}$)

f 194 r8 and $194\frac{8}{9}$

g 1034 r3 and $1034\frac{3}{4}$

h 472 r5 and $472\frac{5}{7}$

3 a 257 r1   **b** 614 r2

c 379 r3   **d** 647 r3

e 2011 r2   **f** 1007 r6

Challenge

1 a 178.25   **b** 163.6

c 453.5   **d** 124.5

e 637.25   **f** 1171.25

g 5730.5   **h** 14 201.6

i 3478.75

---

2 a 634.75   **b** 2484.33

c 1411.17   **d** 7280.75

e 10 149.67   **f** 11 778.86

g 11 279.5   **h** 6041.22

i 3626.43

3 Look for students who show an understanding of real-world numbers by identifying that, although 23 455 ÷ 8 = 2931.875, the actual number of boxes (of eight pencils) is 2931 (with a remainder of seven pencils).

Mastery

1 a The total of the six scores is 317, making the mean average score 52.83.

b The new average score is 317 divided by 7 = 45.29.

c He must score at least 106 in the eighth innings to bring his total to 423. This would raise his average to 52.88. Students may choose to multiply 53 × 8 = 424, raising his new average to 53.

2 Students will need to calculate the week's total as 31 × 7 = 217. The three temperatures shown total 94.5°C. If this is subtracted from 217, it means that the other four days must total 122.5. Look for students who choose realistic temperatures, such as 30.5°C, 30°C, 31°C and 31°C.

3

Country	Amount of potatoes per person per week	Amount of potatoes per person per day
Belarus	3420.41 g	488.63 g
Poland	2616.11 g	373.73 g
Russia	2513.91 g	359.13 g
Ireland	2436.07 g	348.01 g

b It is likely that students will not all arrive at the same answer, as this will depend on the search engine. The date (and reliability) of the data could also affect the answers. The following shows a possible response:

	Country	Potato production
1	China	95 941 500 t
2	India	45 343 600 t
3	Russia	30 199 126 t
4	Ukraine	22 258 600 t

UNIT 1: Topic 10 Integers

Practice

1 a −4 + 5 = 1   **b** 3 − 5 = −2

c 6 − 8 = −2   **d** 0 − 7 = −7

e −6 + 9 = 3   **f** −2 + 7 = 5

g 1 − 7 = −6   **h** 2 + 5 = 7

i 0 + 6 = 6   **j** 7 − 11 = −4

k −7 + 11 = 4   **l** 0 + 8 = 8

2 If students are asked to check with a calculator and it is the kind that does not allow them to enter a negative number, teachers could use this as a discussion point

---

as to why their human brain is more reliable (as in Q 2b) than the electronic "brain" (or the way in which the calculator has been programmed).

a −5   **b** 1   **c** −25

d −70   **e** 25   **f** −3

Challenge

1 a −45, −40, −35, −30, −25, −20, −15, −10, −5, 0, − 5, 10

b −24, −21, −18, −15, −12, −9, −6, −3, 0, 3, 6, 9

c −36, −27, −18, −9, 0, 9, 18, 27, 36, 45, 54, 63

d −200, −175, −150, −125, −100, −75, −50, −25, 0, 25, 50, 75

e −36, −30, −24, −18, −12, −6, 0, 6, 12, 18, 24, 30

2 a Queensland: −11°C and 49.5°C

b −23°C + 8°C = −15°C

c 55.8°

d 73°C (the difference between −23°C and 50°C; New South Wales

Mastery

1 a

Date	Credit $ (Money going in)	Debit $ (Money going out)	Balance $
1 June	500.00		500.00
3 June		125.50	374.50
5 June		69.00	305.50
10 June		420.10	− 114.60
16 June		81.75	− 196.35
21 June	325.00		128.65
30 June		− 348.95	− 220.30

b $−220.30 + $500.00 = $279.70

2 a

Throw	+ or −	Number on dice	Player A moves	Player B moves
1	negative	2	−2	+2
2	positive	3	+3	−3
3	negative	4	−4	+4
4	negative	1	−1	+1
5	positive	6	+6	−6

ANSWERS

b

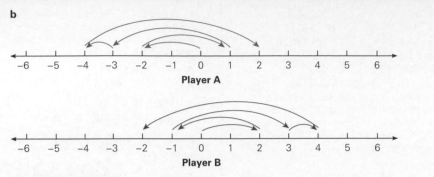

Player A

Player B

c Player A wins by 4.

UNIT 1: Topic 11 Exponents and square roots

Practice

1 a $3 \times 3 \times 3$ **b** $5 \times 5 \times 5 \times 5$ **c** 6×6 **d** $4 \times 4 \times 4 \times 4$

2

	Base number and exponent	Number of times the base number is used in a multiplication	Multiplication	Value of the number
a	7^2	two times	7×7	49
b	8^3	three times	$8 \times 8 \times 8$	512
c	4^2	two times	2×2	4
d	10^5	five times	$10 \times 10 \times 10 \times 10 \times 10$	100 000
e	7^3	three times	$7 \times 7 \times 7$	343

3

	Square number	Exponent	Square root fact
a	36	6^2	$\sqrt{36} = 6$
b	16	4^2	$\sqrt{16} = 4$
c	25	5^2	$\sqrt{25} = 5$
d	49	7^2	$\sqrt{49} = 7$

Challenge

1 a $5^2 = 25$ **b** $5^3 = 125$

2 a $4^2 = 16$ **b** $4^3 = 64$

   c $4^4 = 256$ **d** $4^5 = 1024$

3 a Between 7 m and 8 m

   b Between 5 m and 6 m

   c Between 8 m and 9 m

   d Between 4 m and 5 m

4 121, 196, 289, 400, 484, 529, 625, 1024, 1089, 2116

Mastery

1 A range of answers is possible. Look for students who understand that they need to use 2 as the base number with their exponents. For example:

Day 2: $2^2 = 4$

Day 3: $2^3 = 8$

Day 4: $2^4 = 16$

Day 5: $2^5 = 32$

Day 10: $2^{10} = 1024$

2 If students use perfect square roots, the following options are possible:

4 m² court with 2 m sides; 16 m² court with 4 m sides; 25 m² court with 5 m sides; 36 m² court with 6 m sides; 49 m² court with 7 m sides; 64 m² court with 8 m sides; 81 m² court with 9 m sides; 100 m² court with 10 m sides; 121 m² court with 11 m sides; 144 m² court with 12 m sides; 169 m² court with 13 m sides; 196 m² court with 14 m sides.

Students may choose to use approximate or exact square roots using a calculator for other sizes.

3 Teacher to check. Look for students who show an awareness of an appropriate size – for example that a four-square court that is 14 metres by 14 metres would be too big and one that is 2 metres by 2 metres would be too small.

UNIT 2: Topic 1 Fractions

Practice

1 $\frac{2}{6}, \frac{3}{9}, \frac{4}{12}$

2 Teachers could ask students to justify other answers.

   a $\frac{2}{5}$ **b** $\frac{2}{3}$ **c** $\frac{1}{2}$

   d $\frac{1}{3}$ **e** $\frac{1}{5}$ **f** $\frac{3}{4}$

3 a $\frac{1}{3}$ **b** $\frac{4}{6}$ **c** $\frac{8}{12}$

   d $\frac{2}{3}$ **e** $\frac{1}{6}$ **f** $\frac{3}{4}$

Challenge

1 Teacher to check shading and to decide how accurate the individual students need to be when dividing the shapes.

2 Answers other than those shown below are possible. Teachers could ask students to justify other responses. Likely correct answers are:

   a $\frac{2}{12}, \frac{1}{6}$

   b $\frac{3}{4}$ and any equivalent faction, such as $\frac{6}{8}$

   c $\frac{6}{12}$ of the line is between the two shapes so students could list any fraction that is equivalent to $\frac{1}{2}$.

   d Students will, hopefully, recognise that $\frac{10}{12}$ is an equivalent fraction to $\frac{20}{24}$ and will draw a star above the $\frac{10}{12}$ mark.

e Any fractions that are equivalent to $\frac{20}{24}$ are acceptable. In addition to $\frac{10}{12}$, students will probably identify $\frac{5}{6}$ as an equivalent fraction.

3 a $\frac{2}{4} = \frac{1}{2}$. Numerator and denominator ÷ 2

   b $\frac{3}{12} = \frac{1}{4}$. Numerator and denominator ÷ 3

   c $\frac{2}{10} = \frac{1}{5}$. Numerator and denominator ÷ 2

   d $\frac{8}{12} = \frac{2}{3}$. Numerator and denominator ÷ 4

   e $\frac{5}{6} = \frac{10}{12}$. Numerator and denominator × 2

   f $\frac{6}{8} = \frac{3}{4}$. Numerator and denominator ÷ 2

   g $\frac{4}{5} = \frac{8}{10}$. Numerator and denominator × 2

   h $\frac{6}{9} = \frac{2}{3}$. Numerator and denominator ÷ 3

   i $\frac{3}{5} = \frac{6}{10}$. Numerator and denominator × 2

   j $\frac{15}{20} = \frac{3}{4}$. Numerator and denominator ÷ 5

Mastery

1 a $\frac{1}{5}$ **b** $\frac{1}{4}$ **c** $\frac{3}{4}$

   d $\frac{1}{2}$ **e** $\frac{1}{4}$ **f** $\frac{9}{10}$

2 a There are 8 of the 32 slices left. The lowest equivalent form is $\frac{1}{4}$.

   b 24 of the 32 slices have been eaten. This could be written as $\frac{12}{16}, \frac{6}{8}$ or $\frac{3}{4}$.

3 a Teacher to check explanation. Reducing $\frac{800}{1200}$ to its lowest equivalent form will show that two-thirds of the bricks were broken.

   b The refund should be $\frac{2}{3}$ of \$408 = \$272 ($\frac{1}{3}$ = \$136, so $\frac{2}{3}$ = \$272.)

   c If three-quarters of the bricks had been broken, there would have been 900 broken and 300 unbroken bricks.

4 The easiest solution is to draw a rectangle. Students who show an excellent understanding of equivalent fractions will divide the rectangle into twenty-fourths. $\frac{1}{4}$ of the cake is $\frac{6}{24}$. This leaves $\frac{18}{24}$ to be shared between the six people. Each person will get $\frac{3}{24}$ of the cake. Reduced to its lowest equivalent form, this is $\frac{1}{8}$ of the cake each.

UNIT 2: Topic 2 Adding and subtracting fractions

Practice

1 a $\frac{7}{10}$ **b** $\frac{7}{8}$

   c $\frac{3}{5}$ **d** $\frac{6}{7}$

   e $\frac{9}{10}$ **f** $\frac{4}{5}$

   g $\frac{5}{8}$ **h** $\frac{10}{10}$ or 1 (whole)

   i $\frac{8}{8}$ or 1 (whole)

2 Teachers could encourage students to reduce the answers to their lowest equivalent form where appropriate.

   a $\frac{2}{8}$ (or $\frac{1}{4}$)   **b** $\frac{4}{10}$ (or $\frac{2}{5}$)   **c** $\frac{4}{12}$ (or $\frac{1}{3}$)

   d $\frac{5}{8}$ **e** $\frac{2}{5}$ **f** $\frac{2}{12}$ (or $\frac{1}{6}$)

   g 0   **h** $\frac{13}{20}$ **i** $\frac{10}{15}$ (or $\frac{2}{3}$)

3 Teachers could encourage students to reduce the answers to their lowest equivalent form where appropriate.

   a $\frac{3}{8}$ **b** $\frac{5}{6}$ **c** $\frac{5}{10}$ (or $\frac{1}{2}$)

   d $\frac{7}{8}$ **e** $\frac{7}{9}$ **f** $\frac{5}{12}$

OXFORD UNIVERSITY PRESS

4 Teachers could encourage students to reduce the answers to their lowest equivalent form where appropriate.

a $\frac{5}{10}$ (or $\frac{1}{2}$)   **b** $\frac{2}{10}$ (or $\frac{1}{5}$)   **c** $\frac{1}{12}$

d $\frac{7}{10}$ **e** $\frac{8}{12}$ (or $\frac{2}{3}$)   **f** $\frac{1}{10}$

5 Students should show that $\frac{1}{2} + \frac{3}{8} = \frac{4}{8} + \frac{3}{8} = \frac{7}{8}$

Challenge

Teachers could encourage students to reduce fractions to their lowest equivalent form where appropriate for the activities on this page.

1 a $\frac{3}{8} + \frac{6}{8} = \frac{9}{8} = 1\frac{1}{8}$ **b** $\frac{4}{10} + \frac{9}{10} = \frac{13}{10} = 1\frac{3}{10}$

c $1\frac{6}{8} + \frac{5}{8} = 2\frac{3}{8}$ **d** $2\frac{4}{6} + \frac{5}{6} = 2\frac{9}{6} = 3\frac{3}{6}$

e $3\frac{7}{8} + \frac{6}{8} = 4\frac{5}{8}$ **f** $2\frac{5}{6} + \frac{2}{6} = 2\frac{7}{6} = 3\frac{1}{6}$

2 a $1\frac{1}{8} - \frac{2}{8} = \frac{7}{8}$ **b** $3\frac{7}{10} - \frac{5}{10} = 3\frac{2}{10}$

c $2\frac{3}{12} - \frac{6}{12} = 1\frac{9}{12}$ **d** $1\frac{2}{6} - \frac{5}{6} = \frac{3}{6}$

e $3\frac{1}{4} - 2\frac{2}{4} = \frac{3}{4}$ **f** $1\frac{5}{12} - \frac{8}{12} = \frac{9}{12}$

3 a

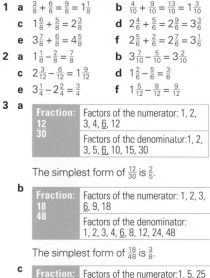

Fraction: $\frac{12}{30}$	Factors of the numerator: 1, 2, 3, 4, 6, 12
	Factors of the denominator:1, 2, 3, 5, 6, 10, 15, 30

The simplest form of $\frac{12}{30}$ is $\frac{2}{5}$.

b

Fraction: $\frac{18}{48}$	Factors of the numerator: 1, 2, 3, 6, 9, 18
	Factors of the denominator: 1, 2, 3, 4, 6, 8, 12, 24, 48

The simplest form of $\frac{18}{48}$ is $\frac{3}{8}$.

c

Fraction: $\frac{25}{45}$	Factors of the numerator:1, 5, 25
	Factors of the denominator:1, 5, 9, 45

The simplest form of $\frac{25}{45}$ is $\frac{5}{9}$.

4 Teacher to check. Look for students who recognise that the overall size of the fractions in the diagram should be the same and who align the equivalent parts of each fraction, such as $\frac{1}{5}$ with $\frac{6}{30}$.

Mastery

1 Students could draw a diagram to show their reasoning. Alternatively, they could use their knowledge of equivalent fractions to show that $\frac{3}{5} = \frac{12}{20}$ and is not, therefore, equivalent to $\frac{15}{20}$.

2 a Teacher to check. Look for students who understand that both the numerators need to be multiplied together and the denominators need to be multiplied together. Although not a requirement of the question, also look for students who reduce the answers to the lowest equivalent fraction.

b Actual expression of the answer will vary. Students should recognise that the numerators are multiplied together and the denominators are multiplied together. The numerator from the original fraction will stay the same as it is multiplied by one while the denominator will quadruple when you are multiplying by $\frac{1}{4}$.

3 a Teacher to check. Look for students who understand the process of finding the reciprocal of the second fraction and multiplying the first fraction by it.

b Actual expression of the answer will vary. Students should recognise that the numerator will double and the denominator will stay the same.

UNIT 2: Topic 3 Decimal fractions

Practice

1 Student should shade:

a any 2 squares   **b** any 25 squares

c any 70 squares   **d** any 45 squares

2 a 0.03 < 0.3: True

b $\frac{1}{1000}$ = 0.001: True

c $\frac{99}{1000}$ = 0.99: False

d 0.03 > $\frac{3}{1000}$: True

e $\frac{735}{1000}$ < 0.735: False

f $\frac{3}{4}$ > 0.075: True

g 1.35 = $1\frac{35}{100}$: True

h 4.350 > 4.35: False

3 Student colours 20 squares blue, 10 squares green, 5 squares yellow, 25 squares red, 39 squares purple = 99 squares in total. The amount left unshaded is $\frac{1}{100}$ or 0.01.

4 0.137, 0.31, 0.37, 0.371, 1.037, 1.37

Challenge

1 Students could be asked to simplify fractions where appropriate. Teachers may wish to discuss the use of the non-significant zero (see question 1f).

	Fraction	Decimal
a	$\frac{9}{1000}$	0.009
b	$\frac{4}{100}$	0.04
c	$\frac{1}{1000}$	0.001
d	$\frac{16}{1000}$	0.016
e	$\frac{1}{4}$ or $\frac{25}{100}$	0.25
f	$\frac{750}{1000}$	0.75 (0)
g	$\frac{99}{1000}$	0.099
h	$\frac{1}{2}$	0.5

2 Teachers may wish to discuss with students the issue of reducing fractions to their lowest forms when dealing with decimals and fractions. For example, 0.8 is more commonly known as $\frac{8}{10}$ instead of $\frac{4}{5}$, whereas $\frac{25}{100}$ is commonly converted to $\frac{1}{4}$.

	Improper fraction	Mixed number	Decimal
a	$\frac{5}{4}$	$1\frac{1}{4}$	1.25
b	$\frac{24}{10}$	$2\frac{4}{10}$	2.4
c	$\frac{175}{100}$	$1\frac{75}{100}$ or $1\frac{3}{4}$	1.75
d	$\frac{250}{100}$	$2\frac{1}{2}$ or $2\frac{50}{100}$	2.5
e	$\frac{475}{100}$	$4\frac{3}{4}$ or $4\frac{75}{100}$	4.75
f	$\frac{2750}{1000}$	$2\frac{3}{4}$ or $2\frac{750}{1000}$	2.75

3 a 0.4   **b** 0.8

c 0.25   **d** 0.75

e 1.4   **f** 1.25

Mastery

1 a 0.625   **b** 0.714   **c** 0.143

d 0.875   **e** 0.583   **f** 0.273

2 a $0.\dot{6}$ **b** $0.1\dot{6}$ **c** $0.1\dot{8}$

d $0.8\dot{3}$ **e** $0.6\dot{3}$ **f** $0.\dot{4}$

3 Students with an excellent understanding should not need to do a great deal of trial and error. They will, hopefully, have discovered that fractions such as fifths and tenths will have no recurring decimals. The easiest solution is to convert thirds, sixths or ninths to decimals.

4 Students may be surprised to learn that many countries use a comma as a separator between whole numbers and decimals. France, Germany and, in fact, most European countries, plus nearly all South American countries, adopt this practice. The UK, USA, Australia and much of Asia use the dot. The internationalisation of learning and the associated need to eliminate confusion is, perhaps, the primary reason we in Australia now use spaces to separate sets of three whole-number digits in the modern counting system.

UNIT 2: Topic 4 Addition and subtraction of decimals

Practice

1 a 4.22   **b** 71.7   **c** 56.165

d 48.1   **e** 1.77   **f** 375.29

g 0.06   **h** 6.57   **i** 135.79

2 a 31.8   **b** 17.65   **c** 48.58

d 8.81   **e** 8.63   **f** 8.6

3 a 28.86 m   **b** 28.925 kg

Challenge

1 9.58 seconds

2 525 mL or 0.525 L

3 The total is 3.62 m so there is a piece that is 28 cm (or 0.28 m) left over.

4 11 servings can be taken (totalling 1485 g, or 1.485 kg).

5 475 g / 0.475 kg

Mastery

1 Students will probably use a process of trial and error to solve the problem. However, look for students who recognise that the last addend must have a 4 in the ones column. The digits in the two blank tenths-columns need to total 7 as there is a 2 in the final tenths-column, and there is a tenth to be traded once the hundredths are added together. Once this step has been reached the gaps can be filled in a logical sequence:

0.1 + 5.63 + 4.27 = 10
0.2 + 5.53 + 4.27 = 10
0.3 + 5.43 + 4.27 = 10
0.4 + 5.33 + 4.27 = 10
0.5 + 5.23 + 4.27 = 10
0.6 + 5.13 + 4.27 = 10
0.7 + 5.03 + 4.27 = 10

ANSWERS

2 Students could be encouraged to solve this problem as part of a cooperative group challenge. Hopefully they will realise that smaller digits will be needed for the whole numbers at positions a, b and c so that the sum will not be too large. (The digit at the d position can be larger.) Look for students who use mental strategies to eliminate some potential solutions before committing pencil to paper. One solution is: a = 2, b = 0, c = 4, d = 8. The addition becomes 2.0 + 4.8 + 2.4 + 0.8 = 10. Some other solutions include:

a = 1   b = 6   c = 0   d = 7
a = 2   b = 3   c = 1   d = 8
a = 2   b = 1   c = 3   d = 8
a = 1   b = 4   c = 2   d = 7

UNIT 2: Topic 5 Multiplication and division of decimals

Practice

1 a 12.3   b 2.34   c 34.5
   d 12.34   e 2.345   f 3.456
   g 456.78   h 87.654   i 98.765
2 a 12.1   b 2.31   c 34.3
   d 4.54   e 5.65   f 6.76
   g 7.87   h 8.98   i 9.109
   j 12.321   k 234.32   l 34.543
3 $4734 ÷ 5 = $946.80. This is $53.20 less than $1000.

Challenge

1 a 8 484.84   b 16 262.61
   c 309.0906   d 309.0906
   e 22 727.25   f 5 090.904
   g 111.0111   h 333.0333
   i 222.0222
2 a $13.33 = $13.35
   b $18.50
   c $21.665 = $21.65
   d $12.49875 = $12.50
   e 2.66 recurring = $2.65
   f $2.85
   g $9.598 = $9.60
   h $33.325 = $33.33 or $33.35 (Possible discussion point as to how to round the number.)

Mastery

1 Five balls will cost $29.35 × 5 = $146.75, so he can afford to buy 4 balls ($29.35 × 4 = $117.40).

2

3	0.1666	2
0.6666	1	1.5
0.5	6	0.3333

UNIT 2: Topic 6 Decimals and powers of 10

Practice

Teachers to decide what is acceptable if students use non-significant zeros in their answers.

1 a 47   b 470   c 4700
2 a 93   b 930   c 9300
3 a 23.9   b 239   c 2390
4 a 30.01   b 300.1   c 3001
5 a 23.42   b 2.342   c 0.2342
6 a 42.3   b 4.23   c 0.423
7 a 723.48   b 72.348   c 7.2348
8 a 26.05   b 2.605   c 0.2605
9 a $235   b 3 tonnes   c 10.5 m
10 a $1.75   b 3.4 tonnes   c 5.242 km

Challenge

1

	× 30	First × 10	then × 3	Multiplication fact
a	1.2	12	36	1.2 × 30 = 36
b	2.3	23	69	2.3 × 30 = 69
c	3.1	31	93	3.1 × 30 = 93
d	1.5	15	45	1.5 × 30 = 45
e	0.32	3.2	9.6	0.32 × 30 = 9.6

2

	× 30	First × 3	then × 10	Multiplication fact
a	1.3	3.9	39	1.3 × 30 = 39
b	2.2	6.6	66	2.2 × 30 = 66
c	3.2	9.6	96	3.2 × 30 = 96
d	5.1	15.3	153	5.1 × 30 = 153
e	0.33	0.99	9.9	0.33 × 30 = 9.9

3 a 68   b 100   c 175
   d 102   e 48   f 250
   g 75   h 144   i 51
   j $60(.00)   k $67.50   l 84 L
   m 249 kg   n 1488 cm or 14.88 m
4 51 kg

Mastery

1 a 70.2   b 57.5   c 75.9
   d 187.5   e 226.8   f 164.5
   g 109.2   h 236.8   i 136.9
2 She has $125.65 × 52 = $6533.80 by the end of the year. This means she is $165.20 short of the price of the car.

UNIT 2: Topic 7 Percentage, fractions and decimals

Practice

1 a 0.04, 4%. Student shades 4 squares.
   b $\frac{23}{100}$, 23%. Student shades 23 squares.
   c $\frac{45}{100}$, 0.45. Student shades 45 squares.
   d 0.4, 40%. Student shades 40 squares.

2 Other equivalent fractions are acceptable.

	Fraction	Decimal	Percentage
a	$\frac{85}{100}$	0.85	85%
b	$\frac{3}{100}$	0.03	3%
c	$\frac{7}{100}$	0.07	7%
d	$\frac{7}{10}$	0.7	70%
e	$\frac{3}{10}$	0.3	30%
f	$\frac{65}{100}$	0.65	65%
g	$\frac{3}{4}$	0.75	75%
h	$\frac{99}{100}$	0.99	99%
i	$\frac{9}{10}$	0.9	90%
j	$\frac{1}{4}$	0.25	25%
k	$\frac{3}{5}$	0.6	60%
l	$\frac{18}{100}$	0.18	18%

3 a $\frac{1}{2}$ = 50%: True
   b 0.07 < 70%: True
   c 0.23 = $\frac{23}{100}$: True
   d 75% > $\frac{3}{4}$: False
   e $\frac{4}{100}$ < 40%: True
   f 0.89 = 89%: True
   g $\frac{1}{5}$ > 20%: False
   h 35% = 0.35: True
   i 1% = 0.01: True
4 a 65%, 0.7, $\frac{3}{4}$
   b 29%, 0.292, $\frac{3}{10}$

Challenge

1 a $\frac{56}{75}$
   b Teacher to check. Look for students who make use of their knowledge of decimals and percentages to make reasonable guesses within an acceptable margin of error of the exact answers.
   c Decimal: 0.746 or rounded to .75. Percentage: 74.67% or rounded to 75%.
2 Teachers to decide the level of tolerance for the responses. Intended positions are:
Percentages:
25%, 40%, 57%, 72%, 99%
Fractions (Allow equivalents):
$\frac{5}{100}$, $\frac{35}{100}$, $\frac{53}{100}$, $\frac{3}{4}$, $\frac{99}{100}$
3 Teachers to decide the level of tolerance for the responses. Positions are:
   a square: 9%
   b star: 40%
   c circle: 58%
   d hexagon: 66%
   e pentagon: 91%
4 Look for students who apply knowledge learned in Unit 2: Topic 3 (Decimal fractions) in order to convert fractions to decimals.
   a square: $\frac{1}{12}$ = 0.08 = 8%
   b star: $\frac{5}{12}$ = 0.42 = 42%
   c circle: $\frac{7}{12}$ = 0.58 = 58%
   d hexagon: $\frac{8}{12}$ = 0.67 = 67%
   e pentagon: $\frac{11}{12}$ = 0.92 = 92%

OXFORD UNIVERSITY PRESS

Mastery

1

	Land area (nearest $\frac{1}{2}$ million km²)	Area which is forest (nearest $\frac{1}{2}$ million km²)	Area which is forest (nearest 1%)	Area of inland water (nearest 5000 km²)	Area which is inland water (nearest 0.1%)
Russia	17 million km²	8 million km²	47%	85 000 km²	0.5%
Brazil	8.5 million km²	5 million km²	59%	59 000 km²	0.7%
Canada	10 million km²	3 million km²	30%	900 000 km²	9%
USA	10 million km²	3 million km²	30%	660 000 km²	6.6%
China	9.5 million km²	2 million km²	21%	270 000 km²	2.8%
Australia	7.5 million km²	1.5 million km²	20%	70 000 km²	0.9%

2 Teachers may wish to discuss why figures such as those in the table are rounded. Actual figures may vary according to the research undertaken.

UNIT 3: Topic 1 Ratios

Practice

1 **a** 5:10   **b** 1:2

2 **a** 20 motorbikes   **b** 200 motorbikes
   c 50 motorbikes

3 **a** 15 trucks   **b** 50 trucks
   c 75 trucks

4 **a** 1:2:3   **b** 15 trucks
   c 40 motorbikes

5 Actual arrangement of colours may vary. There should be 20 red squares, 8 blue squares and 12 green squares.

Challenge

1

	Number of students	Red hair	Blonde hair	Black hair	Brown hair
a	125	10	25	35	55
b	200	16	40	56	88
c	300	24	60	84	132
d	375	30	75	105	165

2 **a** 3:5
   b 2:3
   c 60%

3 **a** 9:25
   b 13:25
   c 3:25
   d 13:9:3
   e 0.12

Mastery

1 Teacher to check. A range of answers is possible. Look for students who can identify and apply a range of ratios to match with the scenario.

2 Teacher to check. A range of answers is possible. Look for students who can generate different responses and correctly calculate the matching fraction, percentage and decimal for both colours.

3 Teacher to check. Look for students who can correctly identify the ratios that match the responses that they generated.

UNIT 4: Topic 1 Geometric and number patterns

Practice

1 Teacher to check, e.g. "Five sticks are needed for every pentagon".

2 **a**

Number of pentagons	1	2	3	4	5	6	7	8	9	10
Number of sticks	5	10	15	20	25	30	35	40	45	50

   b 100

3 Student circles: "Start with one stick and then use four sticks for every pentagon".

4 **a** $5 + 4 \times 2 = 13$
   b $5 + 4 \times 4 = 21$

5 Answers may vary but are likely to be similar to the following:
"You need six sticks for the first hexagon and five for every other hexagon," or "You need one stick to start off and then five sticks for every hexagon".

Challenge

1 **a** Yes: $972 \div 9 = 108$
   b No: $799 \div 9 = 88$ r7
   c Yes: $441 \div 9 = 49$
   d Yes: $4284 \div 9 = 476$
   e No: $3849 \div 9 = 427$ r6
   f Yes: $123\ 723 \div 9 = 13\ 747$

2 Student writes any six different digits, the sum of which is 9, then writes the corresponding division problem. For example, $791\ 235 \div 9 = 87\ 915$.

3 Teacher to check. Correct responses are likely to be similar to the following:

a

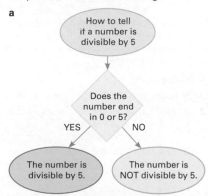

b

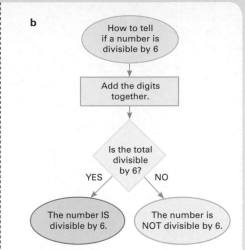

Mastery

1 Look for students who recognise that any number that is divisible by 2 is also divisible by 4 and 8, and do not use them as part of the flow chart. Similarly, numbers that are divisible by 3 are also divisible by 6 and 9. The flow chart is likely to look similar to the following:

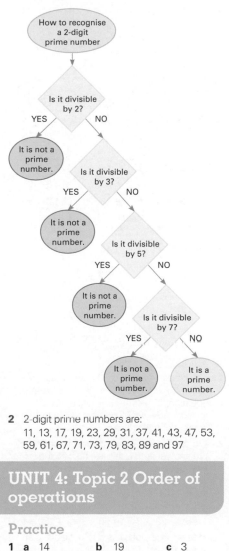

2 2-digit prime numbers are:
11, 13, 17, 19, 23, 29, 31, 37, 41, 43, 47, 53, 59, 61, 67, 71, 73, 79, 83, 89 and 97

UNIT 4: Topic 2 Order of operations

Practice

1 **a** 14   **b** 19   **c** 3
   d 10   **e** 8   **f** 9
   g 21   **h** 14   **i** $7\frac{1}{2}$

ANSWERS

j 6   **k** 27   **l** 81

m 81   **n** 29   **o** 1

p $2\frac{1}{2}$

2

a	42×3	$= (40 \times 3) + (2 \times 3)$	$= 120 + 6$	$= 126$
b	14×6	$= (10 \times 6) + (4 \times 6)$	$= 60 + 24$	$= 84$
c	53×5	$= (50 \times 5) + (3 \times 5)$	$= 250 + 15$	$= 265$
d	23×9	$= (20 \times 9) + (3 \times 9)$	$= 180 + 27$	$= 207$

3 Examples of how the problems might be solved:

a	$25 \times 11 \times 4$	$= 25 \times 4 \times 11$	$= 100 \times 11$	$= 1100$
b	$50 \times 12 \times 4$	$= 50 \times 4 \times 12$	$= 200 \times 12$	$= 2400$
c	$8 \times 15 \times 4$	$= 15 \times 4 \times 8$	$= 60 \times 8$	$= 480$
d	$5 \times 19 \times 2$	$= 5 \times 2 \times 19$	$= 10 \times 19$	$= 190$

Challenge

1

	Problem	Use opposites	Write the value of ◆	Check by writing the equation
a	$\blacklozenge \times 7 = 63$	$\blacklozenge = 63 \div 7$	9	$9 \times 7 = 63$
b	$\blacklozenge - 2\frac{1}{2} = 6$	$\blacklozenge = 6 + 2\frac{1}{2}$	$8\frac{1}{2}$	$8\frac{1}{2} - 2\frac{1}{2} = 6$
c	$\frac{1}{4}$ of $\blacklozenge = 15$	$\blacklozenge = 15 \times 4$	60	$\frac{1}{4}$ of $60 = 15$
d	$\blacklozenge \times 100 = 225$	$\blacklozenge = 225 \div 100$	2.25	$2.25 \times 100 = 225$
e	$\blacklozenge \div 100 = 1.6$	$\blacklozenge = 1.6 \times 100$	160	$160 \div 100 = 1.6$
f	$\blacklozenge \div 8 = \frac{1}{8}$	$\blacklozenge = \frac{1}{8} \times 8$	1	$1 \div 8 = \frac{1}{8}$
g	$\blacklozenge \times 1000 = 7225$	$\blacklozenge = 7225 \div 1000$	7.225	$7.225 \times 1000 = 7225$

2

	Problem	Possible substitutes for ◆				Check by writing the equation
a	$\blacklozenge \times (2 + 3) = 50$	8	9	**10**	11	$10 \times (2 + 3) = 50$ $= 10 \times 5 = 50$
b	$72 \div \blacklozenge - 5 = 3$	**9**	10	11	12	$72 \div 9 - 5 = 3$ $= 8 - 5 = 3$
c	$1\frac{1}{2} \times \blacklozenge + 6$ $= 10\frac{1}{2}$	2	**3**	4	5	$1\frac{1}{2} \times 3 + 6 = 10\frac{1}{2}$ $= 4\frac{1}{2} + 6 = 10\frac{1}{2}$
d	$25 \div \blacklozenge + 2.5$ $= 5$	5	**10**	15	20	$25 \div 10 + 2.5 = 5$ $= 2.5 + 2.5 = 5$
e	$25 \times (10 - \blacklozenge)$ $= 150$	3	**4**	5	6	$25 \times (10 - 4) = 150$ $= 25 \times 6 = 150$
f	$\blacklozenge \times 2 = 3 + 5^2$	12	13	**14**	15	$14 \times 2 = 3 + 25$ $= 14 \times 2 = 28$
g	$(5 + \blacklozenge) \times 10$ $= 25 \times 3$	1.5	2	**2.5**	3	$(5 + 2.5) \times 10$ $= 25 \times 3$ $= 7.5 \times 10 = 75$

3 a $25 \times 4 = 10^2$ **b** $1\frac{1}{2} \times 2 = 3$

    c $4 + 2 \times 3 = 10$ **d** $\frac{1}{2}$ of $6 \times 5 = 15$

Mastery

1 The correct answer can be found using $(27 + 23) \times 3 = 50 \times 3 = 150$. The other equation, $27 + 23 \times 3$, would give an answer of 96, which is incorrect in the context of this problem.

2 Although the students can easily see that the answer is 151, this is a task to see whether they can identify an equation that will solve the problem. Students who are successful in this are likely to be able to apply this skill to other situations.

$1 + (27 + 23) \times 3 = 1 + 50 \times 3 = 1 + 150 = 151$

3 In these questions it is also possible to arrive at the answer without writing equations, but teachers will probably wish to encourage their students to do so in order to improve their problem-solving skills.

a Some students may choose to bracket the numbers being multiplied in order to clarify the equation. Teachers could reassure them that, although this is not necessary, as doing so does not affect the outcome.

$(4 + 2) \times 2 + 3 \times 5 + 2 +$
$2 \times 5 + 2 \times 10$
$= 6 \times 2 + 3 \times 5 + 2 + 2 \times 5 + 2 \times 10$
$= 12 + 15 + 2 + 10 + 20 = 59$
[The first line could be written as $(4 + 2) \times 2 + (3 \times 5) + 2 + (2 \times 5) + (2 \times 10)$ for clarification purposes.]

b The operations can be carried out as follows:
$2 \times 5 + \frac{1}{4}$ of $96 + 2 \times 10$
$= 10 + 24 + 20 = 54$.
This leaves $96 - 54 = 42$ pages left to read.

UNIT 5: Topic 1 Length

Practice

1 a

cm	mm
2 cm	20 mm
3.5 cm	35 mm
4.7 cm	47 mm
12.3 cm	123 mm
15 cm	150 mm

b

m	cm
4 m	400 cm
2.5 m	250 cm
2.32 m	232 cm
4.35 m	435 cm
7.5 m	750 cm

c

km	m
3 km	3000 m
14 km	14 000 m
4.75 km	4750 m
3.9 km	3900 m
1.8 km	1800 m

2 Students could be asked to justify other responses.

	Object	1st unit	2nd unit
a	The length of a paper clip	2.7 cm	27 mm
b	The height of a young student	127 cm	1.27 m
c	The length of a finger	7.5 cm	75 mm
d	The height of a ceiling	2.7 m	2700 mm
e	The length of a street	2.7 km	2700 m

3 a Teacher to check. Most probable answers are rulers, tape measures, trundle wheels and laser tape measures. Students may also suggest non-standard tools such as paper clips.

b Answers will vary. Look for students who can select a tool with an appropriate scale to measure the width of the book and who can use the language of measurement to articulate the reasons for their choice.

Challenge

1 Student should circle 3 m 50 cm and 350 cm.

2 a Estimates may vary, but should be within 1 cm less than 8.5 cm for Line A and 1 cm more than 8.5 cm for Line B.

b Teachers to decide on amount of tolerance for lengths. Suggest +/− 1 mm.
Line A: 8.1 cm
Line C: 9.2 cm

3 a one side   **b** 3 cm

4 Teachers to decide on amount of tolerance for perimeters. Suggest +/− 1 mm per side.

a $2 \times (3.5 \text{ cm} + 1.9 \text{ cm}) = 10.8 \text{ cm}$. Student answers, "I measured two sides".

b $6 \times 1.5 \text{ cm} = 9 \text{ cm}$. Student answers, "I measured one side".

5 Teachers to decide on amount of tolerance for the lengths of the sides and the accuracy of the shape. The aim is for the students to draw a rectangle that is 14.8 cm × 2.6 cm.

Mastery

1 a and **b** Teacher to check. Students could be asked to show their working out.

c Most likely response is 99 cm, but others are possible.

d Answers may vary, e.g. the length of the whiteboard.

2 Students may use other units of length.

a Length: 134 cm − 7.53 cm = 126.47 cm
Width: 99 cm − 33.59 cm = 65.41 cm
Height: 120 cm − 56.5 cm = 63.5 cm

b Practical activity

c Answers may vary, e.g. an armchair.

3 2.58 cm or 25.8 mm

4 a 3000 of the books would fit end to end on a 30 cm ruler (300 mm divided by 0.1 mm = 3000, or 3000 × 0.1 = 300).

b 0.1 mm − 0.03 mm = 0.07 mm. The book is seven-hundredths of a millimetre wide.

5 A micrometre is a unit of length based on the metre. It is one-millionth of a metre. This means that it is one ten-thousandth of a centimetre. There are 1000 micrometres in 1 millimetre. The guitar, therefore, is one-hundredth of a millimetre in length. It has guitar strings which are five-hundredths of a micrometre thick.

UNIT 5: Topic 2 Area

Practice

1 a 18 cm²   **b** 16 cm²   **c** 21 cm²

    d 6 cm²   **e** 12 cm²

2 Balcony: 21 m², bedroom: 18 m², bathroom: 9 m², lounge room: 36 m², kitchen/dining room: 35 m²

Challenge

1 a 14 cm²   **b** 24 cm²   **c** 9 cm²

    d 23 cm²   **e** 42 cm²

2 Teacher to check. Look for students who use the knowledge of splitting shapes into rectangles to create the two different shapes.

Mastery

1 Some students may choose to draw rectangles around the triangles (see Student Book 6, page 93). Others may prefer to imagine the rectangles. Teachers to decide whether students are ready to use a formula for calculating the area of a triangle, e.g. half the length of the base multiplied by the height of the triangle.

a 15 cm²   **b** 6 cm²

OXFORD UNIVERSITY PRESS

2 See also the note for question 1.

   a 24 cm²   **b** 10 cm²

3 Students could spend as little or as much time as they (or the teacher) desire on this task. Students could share their ideas with each other. Alternatively, the task could be completed as a cooperative group activity.

A 9 m × 9 m enclosure would provide the largest area (81 m²). Drawing different rectangles by increasing the length by a metre and reducing the width by a metre each time would use all the wire, but the area would become smaller and smaller (10 m × 8 m gives a perimeter of 36 m but the area reduces to 80 m²; 11 m × 7 m gives a perimeter of 36 m but further reduces the area to 77 m², and so on.).

UNIT 5: Topic 3 Volume and capacity

Practice

1 **a** 18 centimetre cubes **b** 18 cm³

2 **a** 80 cm³   **b** 63 cm³   **c** 84 cm³

3 **a**

Kilolitres	Litres
5 kL	5000 L
2.5 kL	2500 L
4.25 kL	4250 L
3.75 kL	3750 L

    b

Litres	Millilitres
4 L	4000 mL
3.5 L	3500 mL
2.25 L	2250 mL
9.75 L	9750 mL

    c

Volume	Capacity
100 cm³	100 mL
500 cm³	500 mL
175 cm³	175 mL
2000 cm³	2 L

4 375 mL, 600 mL, $\frac{3}{4}$ L, 1 L 200 mL, 1.25 L, 1375 mL, $1\frac{1}{2}$ L

Challenge

1 Teacher to decide on level of accuracy required for the shading. Answers may vary according to the choice of unit of capacity.

   a 1200 mL or 1.2 L or 1 L 200 mL

   b 1250 mL or 1.25 L or 1 L 250 mL

   c 2000 mL or 2 L

   d 1875 mL or 1.875 L or 1 L 875 mL

   e 1975 mL or 1.975 L or 1 L 975 mL

   f 1750 mL or 1.75 L or 1 L 750 mL

2 Answers may vary. Successful responses are likely to include the fact that there are 10 cubes on the bottom layer and three layers altogether, making 30 cubes in total.

3 Likely responses are:
L: 10 cm, W: 3 cm, H: 1 cm; or
L: 15 cm, W: 2 cm, H: 1 cm; or
L: 30 cm, W: 1 cm, H: 1 cm
Teachers may need to discuss with students whether a prism with dimensions of:
L: 3 cm, W: 2 cm, H: 5 cm
is in fact different from one with dimensions of:
L: 5 cm, W: 2 cm, H: 3 cm.

Mastery

1 Teacher to check. Students may include information such as: a megalitre is equivalent to 1 million litres and can be used to measure large bodies of water such as Olympic swimming pools and a gigalitre is equivalent to 1 billion litres and can be used to measure extremely large bodies of water such as oceans. The accuracy of measurements required depends on what the measurements are used for. For example, it may be important to accurately calculate the capacity of a pool in megalitres to make sure it is a standard size and so that enough water can be provided to fill it. The capacity of an ocean would be harder to accurately measure.

2 This could form part of a group or class challenge to find as many different rectangular prisms as possible, each with a volume of 24 cm³. Teachers may wish to discuss with students whether a prism in a different orientation (e.g. using a different face for the base) actually constitutes a different rectangular prism.
Responses should include (in various orientations):
L: 4 cm, W: 2 cm, H: 3 cm
L: 6 cm, W: 2 cm, H: 2 cm
L: 12 cm, W: 2 cm, H: 1 cm
L: 24 cm, W: 1 cm, H: 1 cm
L: 8 cm, W: 3 cm, H: 1 cm
Note: The responses do not include dimensions of prisms that have edges measured in fractions of a centimetre, but this could be an interesting discussion point for students.

UNIT 5: Topic 4 Mass

Practice

1 **a**

Tonnes	Kilograms
2 t	2000 kg
3.5 t	3500 kg
4.25 t	4250 kg
5.175 t	5175 kg
0.975 t	975 kg

    b

Kilograms	Grams
5 kg	5000 g
3.5 kg	3500 g
0.75 kg	750 g
0.45 kg	450 g
3.07 kg	3070 g

    c

Grams	Milligrams
3 g	3000 mg
2.25 g	2250 mg
1.5 mg	1500 mg
0.735 g	735 mg
0.001 g	1 mg

2 Other ways of writing the masses are possible (e.g. using fractions).

   a 3 kg 750 g or 3.75 kg

   b 750 g or 0.75 kg

   c 2 kg 300 g or 2.3 kg

3 Students could be asked to justify an alternative response.

   a B   **b** A   **c** C   **d** B or C

4 Teacher to decide on level of accuracy. There are 250 g increments on the scale, so the pointer should be closer to 4 kg 500 g than to 4 kg 250 g.

Challenge

1 Estimates should be around the ones shown. Teacher to decide on level of accuracy required. However, different ways of writing the mass should correspond with each other, e.g.

   a 2.7 kg, 2 kg 700 g, 2700 g, $2\frac{7}{10}$ kg

   b 870 g   **c** 1.225 kg

2 **a** C, B, A

   b Any mass between the 1300 kg and 1305 kg is correct but units should correspond to each other. Possible answers (without fractions of a kilogram):
1.301 t, 1 t 301 kg, 1301 kg
1.302 t, 1 t 302 kg, 1302 kg
1.303 t, 1 t 303 kg, 1303 kg
1.304 t, 1 t 304 kg, 1304 kg

3 Practical activity. It is doubtful that, unless the mass of each object is recorded on it, the students will estimate to exactly 1 kg. The main aim is for students to show their proficiency in estimating mass.

Mastery

1 **a** 330 kg + 570 kg + 0.62 t + 1.15 t = 2.67 t = Truck C

   b 0.475 t + 1.45 t + 255 kg + 345 kg = 2.525 t = Truck B

   c 0.075 t + 0.4 t + 1 t 425 kg + 900 kg = 2.8 t = Truck A

2 **a** 9.5 kg, 9 kg 500 g, 9500 g

   b 9500 g ÷ 12 = 792 g (to nearest gram)

   c Look for students who work through the problem logically and who use appropriate masses. An easy, although not a very likely, real-life solution would be to start at 791 and subtract/add one each time up to the 11th apple.
791 + 790 + 792 + 789 + 793 + 788 + 794 + 787 + 795 + 786 + 796 = 8701 g
9500 g − 8701 g = 799 g for the 12th apple.

ANSWERS

UNIT 5: Topic 5 Timetables and timelines

Practice

1 a Clock to show 4:34, 16:34   **b** 2:34 am, 02:34   **c** Clock to show 1:13, 1:13 pm
   d 7:23 am, 07:23   **e** Clock to show 11:47, 11:47   **f** Clock to show 12:03, 12:03 am

2

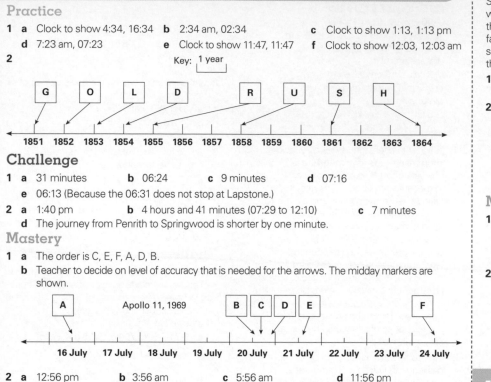

Key: 1 year

Challenge

1 a 31 minutes   **b** 06:24   **c** 9 minutes   **d** 07:16
   e 06:13 (Because the 06:31 does not stop at Lapstone.)

2 a 1:40 pm   **b** 4 hours and 41 minutes (07:29 to 12:10)   **c** 7 minutes
   d The journey from Penrith to Springwood is shorter by one minute.

Mastery

1 a The order is C, E, F, A, D, B.
   b Teacher to decide on level of accuracy that is needed for the arrows. The midday markers are shown.

	Apollo 11, 1969						
A		B C D E					F

16 July   17 July   18 July   19 July   20 July   21 July   22 July   23 July   24 July

2 a 12:56 pm   **b** 3:56 am   **c** 5:56 am   **d** 11:56 pm

UNIT 6: Topic 1 2D shapes

Practice

1 Regular shapes to have horizontal stripes. Irregular shapes to have vertical stripes.
   a irregular triangle
   b regular pentagon
   c irregular quadrilateral
   d irregular octagon
   e regular heptagon (septagon)
   f irregular hexagon

2 Teacher to check. Look for students who note that, because the shape does not have straight sides, it cannot be called a polygon.

3 Students circle, "It is an irregular quadrilateral with one pair of parallel sides".

4 Students draw a square.

Challenge

1 Answers may vary, but look for students who describe the triangle's properties rather than, say, its size.
   a equilateral (All three sides are of equal length. All three angles are the same size.)
   b right-angled (The three sides are different lengths. It has one right angle.)
   c isosceles (Two of the sides and two of the angles are the same size.)
   d scalene (None of the sides and none of the angles are the same size.)
   e isosceles and right-angled (Two of the sides and two of the angles are the same size. It has one right angle)

2 Answers will vary, as any one of the shapes could be "the odd one out". However, look for students who justify their choice(s) by virtue of each shape's properties rather than its appearance. Responses could include, but are not limited to:
   a This is the only shape with two pairs of parallel sides.
   b This is the only shape with two equal sides and two equal angles.
   c This is only shape that is not a polygon.
   d This is the only regular polygon.
   e This is the only polygon with reflex angles.

3 Teachers to decide on level of accuracy. Students should draw a parallelogram with two sides 9 cm long.

Mastery

1 Teacher to decide on level of accuracy required, and to check that students label the circle correctly.

2 Practical activity.

UNIT 6: Topic 2 3D shapes

Practice

1 a cube, prism
   b sphere, other 3D shape
   c hexagonal pyramid, pyramid
   d rectangular prism, prism
   e cone, other 3D shape
   f hexagonal prism, prism

2

	Shape of the base(s)	Number of bases
a	triangular	1
b	triangular	2
c	square	2
d	pentagonal	2

Challenge

Some students will need extra paper on which to try out their sketches. It is important that students do not feel that they have failed if they cannot master the skill of sketching a 3D shape at the first, or even the fifth, attempt.

1 Teachers to decide on the level of accuracy required.

2 Teachers will note that there is no indication of the size of the shapes. The main aim of this task is for students to attempt a representation of the shape. Teachers to use their professional judgement about the degree of accuracy and proportion that is expected.

Mastery

1 Teacher to check. Look for students who show an understanding of the properties of 3D shapes that they can apply to a real-life problem solving situation.

2 Students could practise by making and cutting a cross-section using modelling clay. Others will be able to visualise the cross-section in the abstract. Students should draw:
   a a rectangle   **b** a circle
   c a hexagon   **d** a pentagon
   e a circle   **f** a square

UNIT 7: Topic 1 Angles

Practice

1 a 70°, acute
   b 110°, obtuse
   c 155°, obtuse

2 We know that the angle is 330° degrees, because 360° − 30° = 330°.

3 a 310°   **b** 325°
   c 345°

Challenge

1 Suggested tolerance is +/− 1°.
   a 122°
   b 78°

2 Teacher to decide on desired level of accuracy. Suggested tolerance is +/− 2°.

3 a 85°
   b 117°
   c 88°
   d 322°
   e 124°, 56°

Mastery

1 Teachers may wish to allow a tolerance of +/− 1° but corresponding, supplementary and opposite angles must reflect the size of the initial angle.
   Angle a: 123°
   Angles c, f, h, i, k n, p: 123°
   Angles b, d, e, g, j, l, m, o: 57°

2 Students may wish to carry out research before commencing. The way to work out the sum of the angles in a polygon is to start with the sum of the angles in a triangle, which we already know to be 180°. Each time an extra side is added, we simply add 180°.

So, quadrilateral = 360°, pentagon = 540°, hexagon = 720° and so on.

If students draw irregular polygons (and probably even if they draw regular ones), they may need to "tweak" the measurements of the angle sizes in order for the pattern to be maintained.

UNIT 8: Topic 1 Transformations

Practice

1

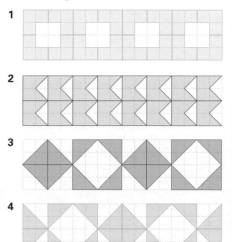

2   **a**   rotation

     b   reflection

     c   translation

3   Student circles, "The shape could have been translated, reflected or rotated".

Challenge

1

2

3

4

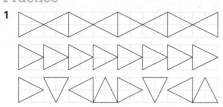

Mastery

1   This could become a "think, pair, share" activity, with students identifying features, such as the parting in the hair, that make the first picture asymmetrical.

2   Teachers to decide on level of accuracy that is required for this difficult activity. The image could be duplicated for students who would like more than one attempt at this task.

3   Practical activity

UNIT 8: Topic 2 The Cartesian coordinate system

Practice

1   **a**   (0,7)   **b**   (−8,1)   **c**   (−5,−8)

     d   (5,−8)   **e**   (8,1)

---

2   The finishing point must be the same as the starting point in order to close the shape.

(−8, 9) → (−5,9) → (−4, 7) → (−5,5) → (−8,5) → (−9,7) → (−8,9)

3   Students join the dots to make a pentagon. Answers for drawing the pentagon could vary depending on the starting point. However, the finishing point must be the same as the starting point in order to close the shape.

(0,7) → (−8,1) → (−5,−8) → (5,−8) → (8,1) → (0,7)

4   Students draw an isosceles triangle of their choice. Answers for drawing the triangle will vary depending on the points chosen. Teachers could ask students to follow each other's instructions to check whether the points have been plotted accurately.

Challenge

1   **a**   (−9, 6) → (−7,7) → (−5,7) → (−4,8) → (−4,6) → (−9,6)

     b   See illustration below for answer.

     c   (9, 6) → (7,7) → (5,7) → (4,8) → (4,6) → (9,6)

2   See illustration below for answer.

3   **a**

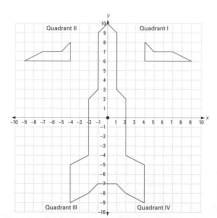

---

b   (0,10) → (1,9) → (1,3) → (2,2) → (2,−4) → (4,−5) → (4,−9) → (2,−8) → (1,−7) → (0,−7)

Mastery

1

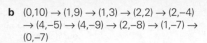

x	0	1	2	3	4	5
y	10	8	6	4	2	0

2   **a**

x	0	1	2	3	4	5	6	7	8	9	10
y	10	8	6	4	2	0	−2	−4	−6	−8	−10

     b   Quadrant IV

     c   See illustration to the right.

3   **a**

x	0	1	2	3	4	5	6	7	8	9	10
y	−10	−8	−6	−4	−2	0	2	4	6	8	10

     b   See illustration to the right.

     c

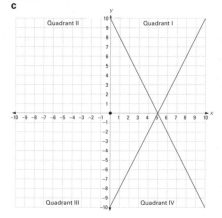

UNIT 9: Topic 1 Collecting, representing and interpreting data

Practice

1   **a**

Frequency table: The number of people who came to visit the school garden				
Day	Monday	Tuesday	Wednesday	Thursday	Friday
Number	11	14	21	22	29

Key ✤ = 4 people

Monday   Tuesday   Wednesday   Thursday   Friday

     b   97

ANSWERS

2 a–b

Graph to show the number of visitors to our garden in open week

c Answers may vary, but should reflect the fact that the number of visitors grew day by day.

d This could form part of a class or group discussion. Some might suggest that the number would be greater than 29 because of the trend, whereas others might suggest that the sixth day would be a Saturday and that there would perhaps be fewer visitors.

Challenge

1 a China

b Answers will vary. Look for students who use their acquired skills to interpret the data, e.g. "The population of the USA is greater than the combined populations of Russia, France and Australia".

2 Answers may vary but are likely to reflect the fact that the approximate size of the populations can be identified on the bar graph.

b 1 400 000 000 (1.4 billion)

c Estimates need to be around 300 000 000. (Data used for the graph was 323 million.)

3 Look for students who use their acquired skills to interpret the data. Answers will vary, e.g. "The pet cat population of the USA is almost half the total pet cat population of all five countries".

4 a The statement is reasonably accurate because 75 million is a quarter of 300 million.

Actual data (rounded at the time of writing) used for the graphs was:

Country	Cats	People
USA	76 million	323 million
China	53 million	1400 million
Russia	13 million	144 million
France	10 million	67 million
Australia	3 million	25 million

Mastery

1 When students are constructing their own graphs, especially when large numbers are concerned, it is important that they are comfortable with rounding to an appropriate number and using a suitable scale for the axes. Teachers may wish to encourage students who undertake any of the activities on this page to discuss the outline of their graphs before they start.

Depending on the experience and ability levels of the students, using digital media to produce graphs can be highly beneficial.

a The gold production for Russia was rounded to the nearest thousand.

b Answers may vary, but are likely to be along the lines of "to save space and/or time". Note that the key to the chart indicated that the number had been abbreviated.

If students are entering large numbers into graphing programs, they will find it easier to abbreviate the numbers (by the same amount) for the data. For example, entering:

China	362
Australia	258
USA	234
Russia	199
South Africa	181

instead of:

China	362 000
Australia	258 000
USA	234 000
Russia	199 000
South Africa	181 000

would result in the same proportions on the graph itself.

2 Practical activity

Practice

1 a Answer should be around 75%.

b 2

2 primary

3 Answers may vary. Look for appropriate responses, e.g. "Do you think that we should be allowed to use our mobile phones in class?"

4 a "Should Year 6 students be allowed to use smartphones in classrooms?"

b Teacher to check reasoning. Students could share their views with each other.

5 Answers may vary but the total of the two should be 100%. Intended response is 45% = No and 55% = Yes, but answers around these figures are acceptable.

Challenge

1 Teacher to check. Students could discuss this question in groups. Likely answer is that the question is misleading.

2 Teacher to check. Students could discuss this question in groups. Intended answer is that it is "based on fact and could possibly be true".

3 Sample survey

4 Answers may vary. Students could discuss this question in groups. Answers may range from, "It is not fair because it doesn't say how many students were surveyed", to "It is fair because it sounds as though it could be true".

5 a Answers may vary, but should reflect an overwhelming No response. Intended response is 3:99 (Yes: No).

b The survey question uses the word "play" whereas the graph title has the word "use". This was not a fair representation of the data.

Mastery

1–2 This page could form the basis of small- or large-group discussion. Answers will vary, and it could be useful for students to write down their ideas and opinions and share them with each other.

3 a The actual percentage was 18%.

b There has been a marked increase in smartphone ownership. Students could be encouraged to share their ideas and opinions with each other.

c A reasonable estimate would be over 90%, but it is doubtful that a figure of 100% would ever be reached.

d Teacher to decide whether this should take place. This could form part of a cooperative group activity.

OXFORD UNIVERSITY PRESS

UNIT 9: Topic 3 Range, mode, median and mean

Practice

1 Range – 82; Mode – 72; Median – 47; Mean – 49

2 a Students' own answers. The mean is the greatest, but accept any answer if students have sound reasoning behind it.

  b 35

  c 49

  d 35

  e Teacher to check written response. Range is 43. Look for students who can accurately define what a range is and who show proficiency in calculating ranges with the given data.

Challenge

1 a 7   b 25

  c 100   d 42

  e 10   f 6

  g 22 or 8   h 121 or 17

2 a 25.125 m   b $25\frac{1}{2}$ m

  c 29 m   d 21.25 m

  e 7 m   f 26 m

3 a 21.4   b 21.5

  c 21   d 12

  e 20

Mastery

1 a Teacher to check. A range of answers is possible. Look for students who understand that the number 12 should be the most common score in their lists of 10 numbers.

  b Teacher to check. A range of answers is possible, but the fifth and sixth largest scores must either both be 10 or must have a mean of 10 – for example, 9 and 11 or 8 and 12.

  c Teacher to check. A range of answers is possible. Look for students who understand that the total of Isabelle's scores must be 100 so that the median will be 10.

  d Teacher to check. The answers will depend on students' responses to questions 1 to 3.

UNIT 10: Topic 1 Describing probabilities

Practice

1 a 1 in 8   b 25%   c 0.75
  d 12.5%   e red

2 Answers will vary. Teachers might wish to ask students to justify their responses.

3 Students circle $\frac{1}{5}$, 20%, 0.2, 1 in 5, $\frac{2}{10}$.

4 Answers may vary but should be appropriate and are likely to be between 75% and 90%. Students could be asked to justify other responses.

Challenge

1 a Students colour the following numbers of sectors:
  • green: 4   • red: 6   • purple: 1
  • yellow: 4   • blue: 3

  b There should be two white sectors out of twenty. These should be described as $\frac{1}{10}$ (or $\frac{2}{20}$), 0.1, 10%.

2 a $\frac{2}{3}$ b 75%   c $\frac{1}{5}$
  d $\frac{7}{12}$ e 0.25   f 0.3̇3̇

3 Students could justify other answers but, based on probability values, the following are likely outcomes. Look for students who apply their knowledge of equivalent fractions.

Bag	Fractions red: green	My prediction after 50 have been taken out	
A	$\frac{4}{5} \cdot \frac{1}{5}$	Red: 40	Green: 10
B	$\frac{1}{10} \cdot \frac{9}{10}$	Red: 5	Green: 45
C	$\frac{3}{10} \cdot \frac{7}{10}$	Red: 15	Green: 35
D	$\frac{3}{5} \cdot \frac{2}{5}$	Red: 30	Green: 20

Mastery

1 Teachers (and students) to decide whether fractions, decimals or percentages should be used for the values. In most cases fractions will be most appropriate because of the total of 52. Equivalent decimals, fractions or percentages are acceptable.

  a 50%

  b $\frac{1}{13}$

  c $\frac{20}{52} = \frac{10}{26} = \frac{5}{13}$

  d $\frac{1}{26}$

  e $\frac{1}{4}$ or 25%

  f $\frac{12}{52} = \frac{6}{26} = \frac{3}{13}$

  g $\frac{8}{52} = \frac{4}{26} = \frac{2}{13}$

  h $\frac{16}{52} = \frac{8}{26} = \frac{4}{13}$

2 a 1 in 13

  b As there are 16 cards worth 10 and there are 51 cards left, the chance is 16 out of 51.

  c $\frac{16}{51}$ = 31% (to the nearest whole number).

3 Depending on the interest level of the students (and teacher) this activity could be as short or as long as desired. Students could, for example, research the probability value of choosing two particular cards by searching online.

UNIT 10: Topic 2 Conducting chance experiments and analysing outcomes

Practice

1 a 25%   b 0.5   c $\frac{3}{4}$

2 a 5   b $\frac{1}{4}$

  c Practical activity. Teachers could encourage students to record the fractions in their lowest equivalent forms.

d Teacher to check. Students are likely to comment that the difference between the prediction and reality is because of the "element of chance" in the game.

3 a Predictions are likely to be half-and-half for each type.

  b Practical activity. Teachers could encourage students to record the fractions in their lowest equivalent forms.

  c As there are only two possible outcomes, their predictions are likely to be closer than in question 2.

Challenge

1 Students may choose to express the probability value as a fraction, a decimal or a percentage. Equivalents of the following are acceptable:

  a $\frac{4}{5}$

  b $\frac{1}{5}$

  c $\frac{2}{5}$

2 a • a queen: 16   • a black queen: 8
      • Jack of Diamonds: 4

  b Practical activity. Teachers could encourage students to record the fractions in their lowest equivalent forms.

  c Because of the chance factor, results are likely to vary from student to student.

3 Because there are 10 cards, the percentages should be easy for students to calculate:

  a 40%

  b 30%

  c 20%

  d 10%

  e 50%

  f 20%

4 a lower ($\frac{4}{11}$)

  b lower ($\frac{3}{11}$)

  c lower ($\frac{2}{11}$)

  d higher ($\frac{2}{11}$)

  e higher ($\frac{6}{11}$)

  f lower ($\frac{1}{11}$)

5 a $\frac{4}{11}$ = 0.36 = 36%

  b $\frac{3}{11}$ = 0.27 = 27%

  c $\frac{2}{11}$ = 0.18 = 18%

  d $\frac{2}{11}$ = 0.18 = 18%

  e $\frac{6}{11}$ = 0.55 (accept 0.54) = 55% (54%)

  f $\frac{1}{11}$ = 0.09 = 9%

Mastery

1 Practical activity. Teachers may wish students to discuss the rules for the game after playing it a few times in order to make it more difficult, more simple or more interesting.

2 This would make a good cooperative group activity.